SKYFARING

A Journey with a Pilot

MARK VANHOENACKER

"A love letter to flight. . . . Vanhoenacker slips easily between poetic meditation on the nature of travel and technical explanations of the mechanisms of the 747, and I found all of it fascinating. It is a delight to encounter someone so unabashedly enamored of the romance of his profession."
—Emily St. John Mandel, *The Millions*

"Vanhoenacker's passionate and beautifully written book will remind even the most jaded traveller of the wonder of flight."
—*The Sunday Times* (London)

"Vanhoenacker is a 747 pilot with a poetic streak. . . . The writing makes flying feel as amazing as it really is."   —Wired.com

"A description of what it's like to fly by a commercial pilot who is also a master prose stylist and a deeply sensitive human being. . . . This couldn't be more highly recommended."
—Alain de Botton, author of *How Proust Can Change Your Life*

"[*Skyfaring*] never loses sight of how beautiful it is to soar above the clouds. . . . [Vanhoenacker's] writing is fluid and elegant."
—*New Statesman* (London)

"Vanhoenacker makes [flying] wondrous again."
—*London Evening Standard*

"Refreshing. . . . A deeply personal and reflective work."
—Patrick Smith, author of *Cockpit Confidential*

Mark Vanhoenacker

SKYFARING

Mark Vanhoenacker is a pilot and writer. A regular con-
tributor to *The New York Times* and *Slate,* he has also written for *Wired,* the *Financial Times,* the *Los Angeles Times,*
and *The Independent.* Born in Massachusetts, he trained
as a historian and worked as a management consultant
before starting his flight training in Britain in 2001. His
airline career began in 2003. He now flies the Boeing
747 from London to major cities around the world.

www.skyfaring.com

SKYFARING

A Journey with a Pilot

Mark Vanhoenacker

Vintage Departures
Vintage Books
A Division of Penguin Random House LLC
New York

FIRST VINTAGE DEPARTURES EDITION, MAY 2016

Two brief quotes are adapted from "With This View, Who Needs Legroom?,"
which appeared on NYTimes.com (March 20, 2010).

The Library of Congress has cataloged the Knopf edition as follows:
Vanhoenacker, Mark.
Skyfaring : a journey with a pilot / Mark Vanhoenacker.
pages cm
1. Airplanes—Piloting—Popular works.   2. Aeronautics—Popular works.
I. Title.
TL710.v345 2015   629.132'52—dc23   2014041159

Vintage Books Trade Paperback ISBN: 978-0-8041-6971-4
eBook ISBN: 978-0-385-35182-9

Author photo © Michael Lionstar

www.vintagebooks.com

Printed in the United States of America
10 9 8 7 6 5 4 3 2 1

For Lois and Mark, and in memory of my parents

. . . Here, as everywhere else,
it is the same age. In cities, in settlements of mud,
light has never had epochs. Near the rusty harbor
around Port of Spain bright suburbs fade into words—
Maraval, Diego Martin—the highways long as regrets,
and steeples so tiny you couldn't hear their bells,
nor the sharp exclamation of whitewashed minarets
from green villages. The lowering window resounds
over pages of earth, the canefields set in stanzas.
Skimming over an ocher swamp like a fast cloud of egrets
are nouns that find their branches as simply as birds.
It comes too fast, this shelving sense of home—
canes rushing the wing, a fence; a world that still stands as
the trundling tires keep shaking and shaking the heart.

—Derek Walcott

Contents

Author's Note

I occasionally struggled to decide which units and terms to use in this book, as aviation itself, though otherwise so globalized, is not always consistent. For example, feet are used to measure height or altitude over most, but not all, of the world—whether or not the metric system is used by those on the ground below. Winds are usually quoted in knots—nautical miles per hour—except where they are quoted in meters per second. My own background as an American working in Britain didn't make things any easier. In general I have tried to use either customary U.S. units or the units most commonly used in aviation. When it comes to talking about the weight—or, more precisely, the mass—of aircraft, though, I have stuck with the metric tons I'm most familiar with at work, as a metric ton (equivalent to around 2,200 pounds) is not so different from a U.S. ton (2,000 pounds).

If you have a favorite photograph from the window seat, please send it along to me via the website Skyfaring.com. I would love to see it.

London
October 2014

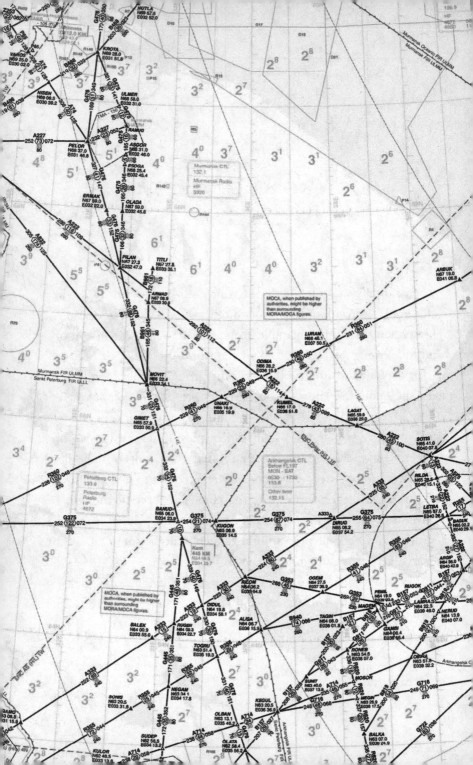

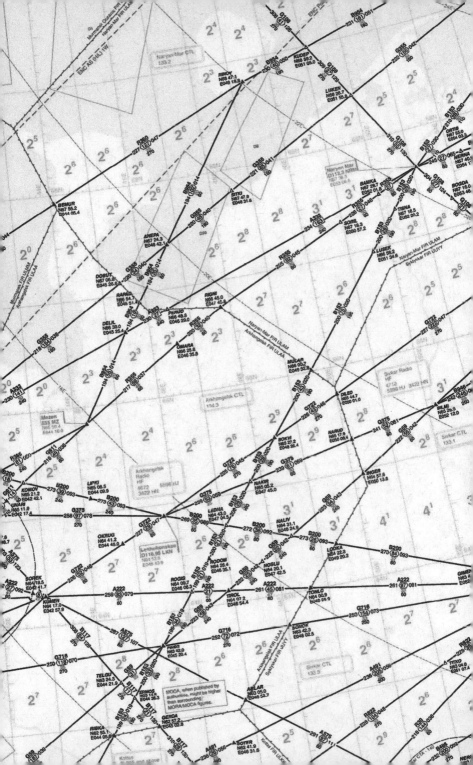

Lift

I've been asleep in a small, windowless room, a room so dark it's as if I'm below the waterline of a ship. My head is near the wall. Through the wall comes the sound of steady rushing, the sense of numberless particles slipping past, as water rounds a stone in a stream, but faster and more smoothly, as if the vessel parts its medium without touch.

I'm alone. I'm in a blue sleeping bag, in blue pajamas that I unwrapped on Christmas morning several years ago and many thousands of miles from here. There is a gentle swell to the room, a rhythm of rolling. The wall of the room is curved; it rises and bends up over the narrow bed. It is the hull of a 747.

When someone I've just met at a dinner or a party learns that I'm a pilot, he or she often asks me about my work. These questions typically relate to a technical aspect of airplanes, or to a view or a noise encountered on a recent flight. Sometimes I'm asked where I fly, and which of these cities I love best.

Three questions come up most often, in language that hardly varies. Is flying something I have always wanted to do? Have I ever seen anything "up there" that I cannot explain? And do I remember my first flight? I like these questions. They seem to have arrived, entirely intact, from a time before flying became ordinary and routine. They suggest that even now, when many of us so regularly leave one place on the earth and cross the high blue to another, we are not nearly as accustomed to flying as we think. These questions remind me that while airplanes have overturned many of our older sensibilities, a deeper part of our imagination lingers and still sparks in the former realm, among ancient, even atavistic, ideas of distance and place, migrations and the sky.

Flight, like any great love, is both a liberation and a return. Isak Dinesen wrote in *Out of Africa:* "In the air you are taken into the full freedom of the three dimensions; after long ages of exile and dreams the homesick heart throws itself into the arms of space." When aviation began, it was worth watching for its own sake; it was entertainment, as it still is for many children on their early encounters with it.

Many of my friends who are pilots describe airplanes as the first thing they loved about the world. When I was a child I used to assemble model airplanes and hang them in my bedroom, under a ceiling scattered with glow-in-the-dark stars, until the day skies were hardly less busy than Heathrow's, and at night the outlines of the dark jets crossed against the indoor constellations. I looked forward to each of my family's occasional airplane trips with an enthusiasm that rarely had much to do with wherever we were going. I spent most of my time at Disney World awaiting the moment we would board again the magical vessel that had brought us there.

At school nearly all my science projects were variations on an aerial theme. I made a hot-air balloon from paper, and sanded wings of balsa wood that jumped excitedly in the slipstream from a hairdryer, as simply as if it were not air but electricity that had been made to flow across them. The first phone call I ever received from someone other than a friend or relative came when I was thirteen. My mom passed me the telephone with a smile, telling me that a vice president from Boeing had asked to speak with me. He had received my letter requesting a videotape of a 747 in flight, to show as part of a science project about that airplane. He was happy to help; he wished only to know whether I wanted my 747 to fly in VHS or Betamax format.

I am the only pilot in my family. But all the same, I feel that imaginatively, at least, airplanes and flying were never far from home. My father was completely enthralled by airplanes—the result of his front-row seat on the portion of the Second World War that took place in the skies above his childhood home in West Flanders. He learned the shapes of the aircraft and the sounds of their engines. "The thousands of planes in the sky were too much competition for my schoolbooks," he later wrote. In the 1950s, he left Belgium to work as a missionary in the Belgian Congo, where he first flew in a small airplane. Then he sailed to Brazil, where in the 1960s he was one of surely not very many priests with a subscription to *Aviation Week* magazine. Finally he flew to America, where he met my mother, went to business school, and worked as a manager in mental health services. Airplanes fill his old notes and slides.

My mother, born under the quieter skies of rural Pennsylvania, worked as a speech therapist and had no particular interest in aviation. Yet I feel she was the one who best understood my attachment to the less tangible joys of flight: the old romance of all journeys, which she gave to my brother and me in the form of stories like *Stuart Little* and *The Hobbit*, but also a sense of what we see from above or far away—the gift, the destination, that flying makes not of a distant place but of our home. Her favorite hymn was "For the Beauty of the Earth," a title, at least, that we agreed might be worth printing on the inside of airplane window blinds.

My brother is not a pilot. His love is not for airplanes but for bicycles. His basement is full of bikes that are works in progress, that he's designing and assembling from far-gathered parts, for me or for a grateful friend. When it comes to his bike frames, he is as obsessed with lightness as any aeronautical engineer. He likes to make and fix bikes even more than he likes to ride them, I think.

If I see my brother working on one of his two-wheeled creations, or notice that he's reading about bikes on his computer while I am next to him on the couch reading about airplanes, I may remember that the Wright brothers were bicycle mechanics, and that their skyfaring skills began with wheels, a heritage that suddenly becomes clear when you look again at their early airplanes. When I see pictures of such planes I think, if I had to assemble anything that looked like this, I would start by calling on the skills of my brother—even though there was the time I got him in trouble with our parents for skipping his chores, and so he taped firecrackers to one of my model airplanes and lit the fuses and waited just the right number of seconds before throwing the model from an upstairs window, in a long arc over the backyard.

As a teenager I took a few flying lessons. I thought that I might one day fly small airplanes as a hobby, on weekend mornings, an aside to some other career. But I don't remember having a clear wish to become an airline pilot. No one at school suggested the career to me. No pilots lived in our neighborhood; I don't know if there were any commercial pilots at all in our small town in western Massachusetts, which was some distance from any major airport. My dad was an example of someone who enjoyed airplanes whenever he encountered them, but who had decided not to make them his life's work. I think the main reason I didn't decide earlier to become a pilot, though, is because I believed that something I wanted so much could never be practical, almost by definition.

In high school I spent my earnings from a paper route and restaurant jobs on summer homestay programs abroad, in Japan and Mexico. After high school I stayed in New England for college but also studied in Belgium, briefly reversing the journey my father had made. After college I went to Britain to study African history, so that I could live in Britain and, I hoped, in Kenya. I left that degree program when I finally realized that I wanted to become a pilot. To repay my student loans and save the money I expected to need for flight training, I took a job in Boston, in the field—management consulting—that I thought would require me to fly most often.

In high school I certainly wanted to see Japan and Mexico, and to study Japanese and Spanish. But really, what attracted me most to such adventures was the scale of the airplane journeys they required. It was the possibility of flight that most drew me to far-off summer travels, to degree programs in two distant lands, to the start of the most literally high-flying career I could find in the business world, and at last—because none of even those endeavors got me airborne nearly often enough—to a career as a pilot.

When I was ready to start my flight training, I decided to return to Britain. I liked many aspects of the country's historic relationship with aviation, its deep tradition of air links with the whole world, and the fact that even some of the shortest flights from Britain are to places so very different from it. And, not least, I liked the idea of living near the good friends I'd made as a postgraduate there.

I began to fly commercially when I was twenty-nine. I first flew the Airbus A320 series airliners, a family of narrow-bodied jets used on short- to medium-distance flights, on routes all around Europe. I'd be woken by an alarm in the 4 a.m. darkness of Helsinki or Warsaw or Bucharest or Istanbul, and there would be a brief bleary moment, in the hotel room whose shape and lay-out I'd already forgotten in the hours since I'd switched off the light, when I'd ask myself if I'd only been dreaming that I became a pilot. Then I would imagine the day of flying ahead, crossing back and forth in the skies of Europe, almost as excitedly as if it was my first day. I now fly a larger airplane, the Boeing 747. On longer flights we carry additional pilots so that each of us can take a legally prescribed break, a time to sleep and dream, perhaps, while Kazakhstan or Brazil or the Sahara rolls steadily under the line of the wing.

Frequent travelers, in the first hours or days of a trip, may be familiar with the experience of jet lag or a hotel wake-up call summoning them from the heart of night journeys they would otherwise have forgotten. Pilots are often woken at unusual points in their sleep cycles and perhaps, too, the anonymity and nearly perfect darkness of the pilot's bunk form a particularly clean slate for the imagination. Whatever the reason, I now associate going to work with dreaming, or at least, with dreams recalled only because I am in the sky.

A chime sounds in the darkness of the 747's bunk. My break is over. I feel for the switch that turns on a pale-yellow beam. I change into my uniform, which has been hanging on a plastic peg for something like 2,000 miles. I open the door that leads from the bunk to the cockpit. Even when I know it's coming—and it's frequently hard to know, depending as it does on the season, the route, the time, and the place—the brightness always catches me off guard. The cockpit beyond the bunk is blasted with a directionless daylight so pure and overwhelming, so alien to the darkness I left it in hours ago and to the gloom of the bunk, that it is like a new sense.

As my eyes adjust, I look forward through the cockpit windows. At this moment it's the light itself, rather than what it falls upon, that is the essential feature of the earth. What the light falls upon is the Sea of Japan, and far across this water, on the snowcapped peaks of the island nation we are approaching. The blueness of the sea is as perfect as the sky it reflects. It is as if we are slowly descending over the surface of a blue star, as if all other blues are to be mined or diluted from this one.

As I move forward in the cockpit to my seat on the right side of it, I think briefly back to the trip I made to Japan as a teenager, about two decades ago, and to the city this plane left only yesterday, though *yesterday* isn't quite the right word for what preceded a night that hardly deserves the name, so quickly was it undone by our high latitudes and eastward speed.

I remember that I had an ordinary morning in the city. I went to the airport in the afternoon. Now that day has turned away into the past, and the city, London, lies well beyond the curve of the planet.

As I fasten my seat belt I remember how we started the engines yesterday. How the sudden and auspicious hush fell in the cockpit as the airflow for the air-conditioning units was diverted; how air alone began to spin the enormous techno-petals of the fans, spin them and spin them, faster and faster, until fuel and fire were added, and each engine woke with a low rumble that grew to a smooth and unmistakable roar—the signature of one of our age's most perfect means of purifying and directing physical power.

In legal terms a journey begins when "an aircraft moves under its own power for the purpose of flight." I remember the aircraft that moved ahead of us for this purpose and lifted ahead of us into the London rain. As that preceding aircraft taxied into position its engines launched rippling gales that raced visibly over the wet runway, as if from some greatly speeded-up video recording of the windswept surface of a pond. When *takeoff thrust* was *set* the engines heaved this water up in huge gusting night-gray cones, new clouds cast briefly skyward.

I remember our own takeoff roll, an experience that repetition hasn't dulled: the unfurling carpet of guiding lights that say *here,* the voice of the controller that says *now;* the sense, in the first seconds after the engines reach their assigned power and we begin to roll forward, that this is only a curious kind of driving down an equally curious road. But with speed comes the transition, the gathering sense that the wheels matter less, and the mechanisms that work on the air—the *control surfaces* on the wings and the tail—more. We feel the airplane's dawning life in the air clearly through the controls, and with each passing second the jet's presence on the ground becomes more incidental to how we direct its motion. Yesterday we were flying on the earth, long before we left it.

On every takeoff there is a speed known as *V1*. Before this speed we have enough room left ahead of us on the runway to stop the takeoff. After this speed we may not. Thus committed to flight, we continued for some time along the ground, gathering still more speed to the vessel. A few long seconds after V1 the jet reached its next milestone of velocity and the captain called: "Rotate." As the lights of the runway started to alternate red and white to indicate its approaching end, as the four rivers of power that summed to nearly a quarter of a million pounds of thrust unfurled over the runway behind us, I lifted the nose.

As if we had only pulled out of a driveway, I turned right, toward Tokyo.

London, then, was on my side of the cockpit. The city grew bigger before it became smaller. From above, still climbing, you realize that this is how a city becomes its own map, how a place becomes whole before your eyes, how from an airplane the idea of a city and the image of a city itself can overlay each other so perfectly that it's no longer possible to distinguish between them. We followed London's river, that led the vessels of a former age from their docks to the world, as far as the North Sea. Then the sea turned, and Denmark, Sweden, Finland passed beneath us, and night fell—the night that both began and ended over Russia. Now I'm in the new day's blue northwest of Japan, waiting for Tokyo to rise as simply as the morning.

I settle myself into my sheepskin-covered seat and my particular position above the planet. I blink in the sun, check the distance of my hands and feet from the controls, put on a headset, adjust the microphone. I say good morning to my colleagues, in the half-ironic sense that long-haul pilots will know well, that means, on a light-scrambling journey, I need a minute to be sure where it is morning, and for whom—whether for me, or the passengers, or the place below us on the earth, or perhaps at our destination. I ask for a cup of tea. My colleagues update me on the hours I was absent; I check the computers, the fuel gauges. Small, steady green digits show our expected landing time in Tokyo, about an hour from now. This is expressed in Greenwich Mean Time. In Greenwich it is still yesterday. Another display shows the remaining nautical miles of flight, a number that drops about one mile every seven seconds. It is counting down to the largest city that has ever existed.

I am occasionally asked if I don't find it boring, to be in the cockpit for so many hours. The truth is I have never been bored. I've sometimes been tired, and often I've wished I were heading home, rather than moving away from home just about as quickly as it is possible to do so. But I've never had the sense that there was any more enjoyable way to spend my working life, that below me existed some other kind of time for which I would trade my hours in the sky.

Most pilots love their work and have wanted to do it for as long as they can remember. Many began their training as soon as they could, often in the military. But when I started my training course in Britain, I was surprised at how many of my fellow trainees had traveled quite far down another path—they were medical students, pharmacists and engineers, who, like me, had decided to return to their first love. For me, coming later to the profession has been an opportunity to think about why many of my colleagues and I were drawn back so strongly to a half-forgotten notion, one that we shared as children.

Some pilots enjoy the hand-to-eye mechanics that are related to movement in three dimensions, particularly the challenges that cluster at the beginning and end of every flight. Others have a natural affinity for machines, and airplanes are engineered nobility, lying well beyond most cars, boats, and motorcycles on the continuum of our shiny creations.

Many pilots, I think, are especially drawn to the freedom of flight. A jet is detached, physically remote and separate for a certain number of miles and hours. Such solitude is all but absent from the world now, and so—paradoxically, for in the cockpit we could hardly be better encased in technology—flight feels increasingly old-fashioned. Paired with this freedom is the opportunity to come to know the cities of the world well and to see so much of the land, water, and air that lie between them.

Then, too, there is the perennial yearning for height that many of us share. High places have gravity. They pull us up. Elevation remains simple, a prime number, an element on the periodic table. "Higher, Orville, higher!" cried the father of the Wright brothers, when he made his first flight at the age of eighty-one. We build skyscrapers and visit their observation decks; we ask for an upper floor in a hotel; we ponder photographs taken from high above our homes, our towns, our planet with a mix of love and bewildered recognition; we climb mountains and try if we can to save our sandwich for the summit. On my first morning in a new city I'll often go first to a viewing point on top of a tall building, where I occasionally see travelers whom I recognize from my flight.

Perhaps evolution alone explains the attraction of altitude. Here is the big picture, the survey, the overview, the lookout, the lay of our land, what approaches our cave or castle. Strabo, the Greek geographer who would partly inspire Columbus, climbed the acropolis of Corinth merely to gain perspective on the city. When my father arrived to work as a missionary in a poor neighborhood of the Brazilian metropolis of Salvador, his first step was to hire a pilot to help him photograph the unmapped neighborhood and its informal, largely unnamed streets. Many years later, after he died, my brother and I heard a rumor that a street in this locality had been named for him after he left Brazil. We pored over a map of the city on a laptop to find Rua Padre José Henrique, Father

Joseph Henry Street; we zoomed in from the digital sky, from four decades and many thousands of miles away, to remember the story of his first flight over this city.

But I think our love of height cannot be entirely explained by its many practical uses. In so many realms we seek evidence of interconnection, of parts that form a whole. In music, comedy, science, we respond to the revealing of relationships we did not see at first, or did not expect to find so pleasing. Flight is the cartographic, planetary equivalent of hearing a song covered by a singer you love, or meeting for the first time a relative whose features or mannerisms are already familiar. We know the song but not like this; we have never met this person and yet we have never in our lives been strangers. Airplanes raise us above the patterns of streets, forests, suburbs, schools, and rivers. The ordinary things we thought we knew become new or more beautiful, and the visible relationships between them on the land, particularly at night, hint at the circuitry of more or less everything.

I've occasionally toured cathedrals in faraway cities that have labyrinths, sinuous paths inlaid in the stone that you follow around and around, back and forth. I've been struck by the peacefulness of labyrinths, the intended result of being able to see your path, and the contrast such a gift makes with the barely relaxing experience of walking a maze, or even the aisles of a supermarket, where you cannot see the whole.

Even today many travelers leave home not just to see new places, but also to see the whole of the place they have left from the various kinds of distance—cultural, physical, linguistic—that travel opens for them. Indeed, a fascination with this perspective is something I associate with the most experienced travelers. Occasionally I fly to a city in which one of the attendants on my flight lives, or was born, and he or she is invariably eager to join us in the cockpit for takeoff or landing, in order to watch how the loved place, though it has no remaining mysteries, leaves the cockpit windows or comes to fill them again.

I love flying, for all these reasons. But to me the joy of airliners is the particular quality of their motion over the world. When I run through the woods, over the ground, the branches are close, loud, fast. I am what's moving. Up and down, turning along the path, my feet never land twice at the same angle. I could stop to touch anything. In contrast, films taken of the earth from orbit show a wholly different kind of motion, a steady and weighty perfection of turning, an imperious stability that's the last thing we might expect from such unfathomable height and speed.

An airliner does not move at either of these extremes. In the course of each flight, however, it crosses much of the continuum between them. I love to fly because I love to watch the world go by. After takeoff we see the world just as we would from a small plane. Then in the high middle hours of a flight we perceive less detail, of course, but we also see a greater extent of the earth than we were surely ever meant to encounter at one time. And in some achingly stately inversion of our senses it's in the cruise, when we are highest and fastest, that place turns most deliberately. The connections below make the most sense to me from this abstracted, apparently slow motion above them. The connections are made as a matter of course, we might say, as a road or a river or a railway runs between two cities, and one landscape or cloudscape flows into another as easily as lines across a page. They also build over time, as the dimensions of a city, a country, or an ocean are summed by the minutes or hours such a place takes to cross the mind's eye.

Then we descend; we *make our approach* to another place. The

world accelerates as we return; it looks fastest just before landing, when the airplane is slowest. The wheels race at takeoff but
are stilled in flight, and on touchdown they are sped up again by
the earth. This touch turns the speed of flight to the speed of the
wheels; the brakes turn this to the heat of home, of a journey's end,
that is carried off on the wind.

A measure of longing is attached to any mode of travel, of
course. By definition every traveler wishes, or needs, to be somewhere else. What is longed for may be the place you have just left.
Or it may be a forest or cathedral or desert you have read about
or imagined since childhood, or a place you have always wished
to live, or a place you knew well when you were young. But flight,
which takes us so far to or from what we love, embodies this longing most directly. The space through which the airplane moves is
so alien. Humans can't breathe in it. We can't pull over halfway
and silence the machine and stretch our legs; we can't swim in it or
hold on to the side of the pool. The adversity of the sky sharply
divides the journey from the times and places that lie at either end.

When travelers move between points on the globe so different
in culture, language, and history—London, Tokyo—the imaginative distance can be as vast as the physical gap in the air above
them. Like the music you love best, this mental distance feels
partly external and partly your own. And so high above the world,
open to more of the planet and sky than any species has the right
to see, we find room for introspection in one of the last places we
might have thought to look for it. When I was thirteen and got
my first portable cassette player and headphones and began to
choose music for myself, I asked my brother if pilots were allowed
to listen to music while they flew. He answered that he wasn't sure,
but he thought not. He was right. But as passengers we are all

given these increasingly rare quiet hours in which there is nowhere we have to go and nothing we have to do, hours in which we are alone with our thoughts and music and the moving picture of our journeys.

Then we blink and suddenly we see again the earth we are flying over. From the window seat our focal point crosses between the personal and the planetary so smoothly that such movement seems to hint at a new species of grace, that we would come to only in the sky. Whatever our idea of the sacred, our simplest questions—how the one relates to the many, how time equates to distance, how the present rests on the past as simply as our lights lie on each night's darkened sphere—are rarely framed as clearly as they are by the oval window of an airplane. We look through it, over snowcapped cordilleras in the last red turn of the day, or upon the shining night-palmistry of cities, and we see that the window is a mirror, briefly raised above the world.

The journey, of course, is not quite the destination. Not even for pilots. Still, we are lucky to live in an age in which many of us, on our busy way to wherever we are going, are given these hours in the high country, when lightness is lent to us, where the volume of our home is opened and a handful of our oldest words—*journey, road, wing, water; earth* and *air, sky* and *night* and *city*—are made new. From airplanes we occasionally look up and are briefly held by the stars or the firmament of blue. But mostly we look down, caught by the sudden gravity of what we've left, and by thoughts of reunion, drifting like clouds over the half-bright world.

Place
✈

I'm thirteen. It's late winter, still bitterly cold. My dad and I have driven from our home in Massachusetts south to New York City. At Kennedy Airport we park on top of the Pan Am terminal. We're here to pick up a cousin of mine who is coming to live with us for a few months. We are early, or perhaps his flight is late. We stay for some time under the gray skies and watch the planes as they ascend from distant runways and roll to the gates beneath us.

Among the coming and going of airliners I see an aircraft from Saudi Arabia approach the terminal. I have loved airplanes since I was a small child, but I feel a new kind of astonishment at this particular plane, at the sword and the palm tree on the tail, and the name on the side of the jet.

For some reason—the day, my age, a sudden new understanding that the cousin who will eat dinner across from me at our table at home tonight is still somewhere in the sky—the sight of this plane mesmerizes me. A few hours ago the jet, and all that it contains, was probably stopping for fuel in Europe, and a few hours before that it was in *Arabia*. When I woke up in my bedroom this morning, when I sat at the table in our kitchen to have my cereal and orange juice, when we got into the car, this aircraft was already hours into a journey that was as routine to it in its realm as my walk to school is in mine. Now my father and I watch the last of its many turns over the earth that day; the plane is parking—what my parents do when they come back from the supermarket, and what a pilot does, too, I realize, even at the end of a journey from a place like Arabia to a city like New York.

The doors and holds of the jet are still sealed. It strikes me that some essence of the day the jet has left behind, the day of some euphonious city name I have read on the globe in my bedroom—Jeddah or Dhahran or Riyadh, surely—might be locked inside. I try to imagine Saudi Arabia, falling back on my limited sense of deserts composed largely of the Saharan sands in *The Little Prince*. The passengers on that plane would fly this far, see from the window the Atlantic pressing on the snowy coast of Canada or New England; and at the same time my dad and I were driving along an icy old parkway through rural New York State, a road that could never connect you to Arabia, except that it runs to an airport and a plane like this one.

The physical achievement of airplanes—that they take us up into the air, that they enable us to fly—is not half their wonder. Place turns before an airplane with perfect steadiness. It appears in the air as our new and gossamer geography of the sky, it passes unseen, behind clouds or within the modern fiction of the flight computers, it flips past so quickly that it is like a conversation overheard in passing, when you cannot gain purchase on any individual word, or even be certain of the language. Then suddenly a pair of wings, this most charmed of our creations, brings us to a new day, a new place, and to such perfect stillness upon it that we are able to step through the unsealed door and start to walk.

*

I am in the cockpit of a 747 over the wintry-white Rockies, which spread out below me to the horizon. The world is divided: blue above, snow below. I remark on how the shadows of the peaks fall on the land; the captain tells me that clockwise is only clockwise because that is the direction of time, of a shadow, on a sundial in the northern hemisphere. A controller speaks to us on the radio, to announce the presence of another aircraft near us, "now at your two o'clock," so we know in which direction to scour the blue. Then she announces our position as it would appear to the other aircraft: at their "ten o'clock." The jet that started at our two o'clock moves to three, then five o'clock, and then it is behind us and we lose sight of it. The hour-places turn like the teeth of gears.

Jet lag results from our rapid motion between time zones, across the lines that we have drawn on the earth that equate light with time, and time with geography. Yet our sense of place is scrambled as easily as our body's circadian rhythms. Because jet lag refers only to a confusion of time, to a difference measured by hours, I call this other feeling *place lag:* the imaginative drag that results from our jet-age displacements over every kind of distance; from the inability of our deep old sense of place to keep up with our airplanes.

Place lag doesn't require the crossing of a time zone. It doesn't even require an airplane. Sometimes I've been in a forest, for a hike or a picnic, and then later the same day I have returned to a city. Surrounded by cars and noise and blocks of concrete and glass, I'll find myself asking, how is it that I was walking in the woods this morning? I know it was only this morning I was in that different place; but already it feels like a week ago.

We evolved to move slowly over the world, in sight of everything en route. It makes sense that passing time and changing surroundings share a rhythm, and that as a consequence further or more different places naturally seem longer ago. The differences between a forest and a city are so enormous that the journey between them interposes itself as a chronological jump, a kind of time-hill.

This is true of all travel; and the greater the contrast the journey draws between home and away, the sooner the trip will feel as if it took place in the distant past. This equation is pushed to its imaginative limit by the airplane, which takes us on journeys almost none of us would ever undertake by other means, to places as different from our home as any on the planet, over many other places we will know only obliquely, if at all.

I sometimes think that there are cities so different in sensibility, culture, and history—Washington and Rio, Tokyo and Salt Lake City—that really they should never be joined by a nonstop flight; that to appreciate the distances between them such a journey should be broken into stages, and that the imaginative distance might be better discerned if such flights took ten weeks, not ten hours. But no matter which pair of cities the plane links, almost all air travel can feel too quick. We pretend that it's normal, that London, the place we were, the place that surrounded us in every respect, has transformed itself into Luanda or Los Angeles, as if it was not us who moved, but rather place that flowed around us, because after all, no one could move this quickly. I listen to Joni Mitchell's "Hejira" and feel "porous with travel fever," porous to the modern fluidity of place.

If we do not see much of the intervening earth—if we as passengers sleep most of the way or do not have a window seat—then journeys of such inconceivable scale can seem to take place all but instantaneously, the airplane door like the shutter of a camera.

It is right that our first hours in a city feel wrong, or at least bewildering, in a way we can't quite specify. We are not built for speed, certainly not for this speed. When we cross the world some lower portion of our brains cannot understand what has, we might say, taken place. I can say matter-of-factly to myself: "I flew from home to Hong Kong. Clearly, this is Hong Kong: the destination signs on the fronts of the buses, the rivers of pedestrians, the surface of the harbor where the lights of so many boats race over the heaving, blurred reflections of skyscrapers." Equally, I know that a day or two ago I was at home. I have the everyday memories, the receipts to prove it. Yet, just as with two disparate times from my own past, I am the connection between these wildly different places across 6,000 miles of intervening continent. Somewhere in my lower-brain unconscious, *I* am the most obvious answer to the question of what these places, separated not by an inconceivable distance but by mere hours, have in common. And that makes no sense at all.

If place lag were a more recognized term, the next time I walked down a street in Tokyo and a van blaring political announcements for a municipal election went past, or I stood in a food market in São Paulo and saw a dozen fruits I did not know how to name or eat, or the skies opened in Lagos and I saw rain the likes of which I would never see in Massachusetts, I could blink and say to my companion, who would nod and smile in recognition: "I have place lag."

For pilots, flight attendants, and the most frequent business travelers, place lag may be a more common experience than jet lag. We rarely stay long enough to adjust to local time—to *acclimatize* (the formal term that appears in regulations specifying the rest a pilot requires after a flight)—before it is time to fly back. I never change my watch or cell phone to local time. Many pilots find it easier to eat and sleep on their home time zone for such short stays, even when very far from it, even when this means a complete reversal of night and day, even if this means three days in a city and never walking through it in daylight.

Place lag, unlike jet lag, may get worse with the passage of time. A huge proportion of our memories relates to the most recent minutes, days, or weeks of our lives. So the first days in a foreign city, even as our bodies begin to adjust to the new time zone, fill our minds with the accumulating incongruities of a new place, displacing the presence and immediacy of our now distant homes. The world gets stranger by the hour.

Travelers may know the experience of arriving in a city late at night, tired and unsure of where to go, and acquiring a specific feeling of the place; then, the next morning, waking in a hotel and opening the curtain to light and life outside the window, and having the sense of arriving anew, or even arriving for the first time, as if what happened at night did not happen at all. When I flew to Delhi for the first time it was January, and the city's famous fog was thick at the airport and in the capital itself. It was perhaps three in the morning when our bus left the terminal. The streets quickly turned narrower, more residential. I was surprised that Delhi that night was far colder than London, and the gray dust on the streets, in the night drifts of fog, looked like nothing so much as snow. In my memory the journey was utterly silent; all I could think of was that we were stealing into Delhi, strangers to the city in both time and place.

Eventually, many passengers will have enough time to replace themselves in this new locale, like a cartoon shadow that's briefly separated from its owner and later reunited. But before this can happen the crew from their flight will almost certainly have gone back to where they came from; we will probably already have flown to yet another city. Equipped with eyeshades and earplugs, and largely free from locally timed schedules in each city we visit, we have more control than most travelers over how much jet lag we experience. But place lag is an unavoidable and all but permanent presence in our lives.

When I have a free morning, I often go to a city's main railway station. New or old, in Beijing or Zurich, the stations are typically masterpieces of architecture, and there are always cafés to linger at with a book. I like, too, the signs on the airport-like departure boards for many smaller cities I have not heard of, or did not realize were close enough to be reached by train. But sometimes I think that the real reason I like to wander or sit in these stations is because they are incarnations of in-betweenness. A busy foreign station looks exactly how I feel.

Place lag is most acute when we depart from a foreign city in the late evening. We board a bus at our hotel and journey to the airport, passing the cars or other buses filled with workers making their late way home, and shopping bags filled with what someone will cook; perhaps they're listening to music or to a sober-voiced news anchor reading out the evening's top stories from what to me might as well be another world. Tonight everyone I see on this road will sleep in their own beds, while I'll be watching the flight instruments and drinking tea over Pakistan or Chad or Greenland. Occasionally on these bus journeys, I experience clarifying jolts of my current place, blasts of the truth that only foreigners will see of a city and a day, the privilege of the outsider's view. But often I feel that I have already left, or that I was never in the city at all.

Later, several hours into a flight, I may think back to the staff we have left behind in Johannesburg or Kuwait or Seattle or Tokyo, those who "work on the ground," as we say, and about the world they return to when their day's or night's tasks are completed, when they disengage from the plane as cleanly as the fueler from the wing. I think about what time it is now, in their city, and whether it's dark yet. I try to imagine what they will eat, or what they will say about their day; what the homes they have gone to look like—Indian or Japanese or American, and each home itself a country.

Although place lag is more a feature of a pilot's life than jet lag, it retains analogies to time. When I see an old black-and-white photograph, I have to remind myself that the world was in color when it was taken; or that to the people in it, the moment captured felt as much like the present as the moment in which I am now looking at the photograph. Place lag is the geographic equivalent of this chronological effect, a dislocation only airplanes are fast enough to conjure from the present moments that run not chronologically down through the past, but horizontally, across the geography of the earth. It's our experience of a truth we could never have evolved to grasp easily: that the whole world, every place, is going on at once.

One winter night I flew to New York, as a passenger. The plane was nearly empty. I was in a middle seat, but the windows on this plane were larger than most and if I sat up straight I had a clear view of the city scrolling past the glowing ellipsis of the windowpanes. In their stowed position the individual passenger television screens faced sideways, out toward the windows and the world.

As we came in to land these unwatched televisions were still on. When I looked toward the windows I saw their images, partially reflected back into the plane. Projected against the night was a comedian at a stand-up club, somewhere and some time else. His glowing, moving image, his silent, laughing audience, rolled smoothly over the turning illuminations of the city. Further down the plane, from another television, a flickering African savanna floated over the sky. Lions turned their faces sideways in inaudible roars and prowled over their unexpected night dominion.

I found myself recalling the memorable name of a category of papal address: *Urbi et Orbi*—to the city and the world. Here we see place more clearly than ever; here we see one city that is given to us so beautifully, that gathers beneath us in the form of its own electrified approximation. Yet here, too, are places crossing places, unmoored and frictionless in the world made by airplanes.

"Twelve hours, twelve bounteous hours are gone, while I / have been a traveler under open sky," wrote Wordsworth. Twelve hours in a 747 is a fair run under the blue or the stars; Tokyo to Chicago, Frankfurt to Rio de Janeiro, Johannesburg to Hong Kong.

I struggle for a means to measure out the human scale of these journeys. The task gets harder, not easier, the more I fly. Sometimes after a long flight I reach my hotel room and close my eyes, and I'm hit by the silence of being alone for the first time in thousands of miles, and I don't know how many faces I've seen since my day began, since the sun rose in whatever city I happened to wake up in that morning. I am certain that on most workdays I see more people than many of my ancestors saw in an entire lifetime. I think how those I've seen have been scattered by the hours of airplanes, how the simplest definition of community, of sharing a space, has been disassembled, even as the plane has enabled new forms of reunions, those that take place on a fully planetary scale. By nightfall many of the people I saw in the airport or onboard my plane will have taken further flights, or will be at home, or in a hotel room like mine. Some may be driving the last miles down a narrow road, completing their journey to a place distant in every sense from the world I know, or may even now be describing their journey to the person they've traveled so far to see.

Sometimes, trying to imagine the dimensions of modern flight, I think of the air. Not just of the volume or depth of air we move through, or Wordsworth's open sky, but rather of their opposite. It's ironic that what's called air travel, which vaults us through so much of the world's air, is so profoundly cut off from any direct physical encounter with it. I suspect this may be the sharpest contrast between those who flew in open cockpits and those who fly now. Who knows what teleportation might feel like; presumably I'll be looking for work as soon as someone finds out. But I imagine we already have a small sense from the air-conditioned boxes and tubes, so well prepared for us, that can convey us nearly anywhere on the planet.

I wake up in a hotel room, after a long postflight nap. I'm in a hotel in an Asian metropolis. It takes me a moment to remember which one. I remember the name of the city just before I sit up, stand, go to the window, draw the curtain back on a harbor filled with moving light, a maritime scene so frenetic it could be a far older age. I lift my gaze, and before looking for the airplanes descending over this waterscape, I pause to look at the noble skyscrapers behind the glittering logos and signs hardly smaller than the faces of the buildings. I shower, dress, wander outside into the electric evening, amid all the light, all the workers rushing home or to meet friends. I look up to where the upper floors of the towers thin out in a starless haze, and I can't calculate how many hours and miles have passed since I was last outside under the open sky.

I skid over the miles and the hours, tripping over the threads that can't be cut, that constitute my various lags. I remember a dark early start in London, a walk to a Tube station, an unconsidered last moment of unmediated sky, when I did not even pause to consider a farewell. Then a train, to another train that took me to the depths of an airport; a walk through the terminal, another underground train, a covered jetway to a plane bound for Hong Kong; a bus from the covered airport station to beneath the large awning of our hotel; automatic doors, banks of shiny elevators with music playing inside and advertisements on the walls for the rooftop jazz lounge; my room and sleep. A journey nearly as momentous as any we can make on the earth; yet not one mile or moment of it under open sky.

The ease with which we cross the world now would certainly shock previous generations. But our ancestors might be equally surprised that it's possible to make such a journey without seeing the sky, or without, at least, the permanent mediation of glass. And air travel is often the most enclosed portion of such journeys. I can enter a terminal in one city and take a series of connecting flights, be carried across the world in no small measure by the wind; I can shop and sleep and dine along the way and yet never face a local breeze.

I often try to open a window in the hotel rooms where I sleep.

In many hotels, none can be opened at all. The term *built environment* typically refers to the totality of man-made features such as streets, parks, and buildings. But one subset of this, the cocoon of glassed-off insulation that is modern travel—in particular, the global house of sealed comfort that air travelers are presumed to want—is a more compelling object for the name.

The completeness of the built environment, the built sky, is often taken as a mark of the quality of the airport, or even of the level of development in a country. Few travelers enjoy boarding a plane that is parked away from the terminal, which may involve waiting on stairs in the wind and the rain. Jetways—or air bridges, a term in which the increasingly sealed-off modern traveler might hear a touch of irony—are added as airports develop and expand. Like aviation itself their shiny presence is taken for a sign of progress.

The extent of the built air is revealed most clearly when it breaks down. Even when the plane is attached to the terminal by a jetway, if the seal it makes is imperfect, where the edge of the climate-controlled jetway meets the plane there are brief little gusts of Dallas heat or Brussels damp or Moscow cold. Such air feels and smells different from the conditioned environment; it hits me like a transgression, but also a blessing of place—a sudden blast of place lag, perhaps, but also the first breath of what will eventually remedy it. Honolulu, with an open-sided, though still covered, terminal, is a rare exception in the world of large airports. I was dumbfounded when I first walked through it, not by the volumes it speaks about Hawaii's weather, but by what was for me the extraordinary sensation of natural, fragrant air washing though the sanitized realm of global aviation.

If the enclosed airspace of the world—"breathing what is called air," in poet W. S. Merwin's description of waiting in an airport's atmosphere—is a sad thing, an effacement of place or a modern excess of insulation and comfort, it has the advantage that it makes arrival in the true air of a city much more vivid. If I sailed from one city to another slowly, exposed to long weeks of the elements, I might not notice how sharp the air differences are between the two places.

Flying into certain Indian cities, I have come to recognize and love the unique and rich, faintly smoky smell that I have been told comes from the burning of biomass and fuel derived from cattle waste. It must rise through, or pool at, certain altitudes. Often I can smell it in the cockpit at night, in the last minutes before we land. Particularly if you are from one of these cities and have been away for a long time, this must be a pleasing thing to recognize, an unmistakable and physical quality of the air that returns in rough symmetry with the lights of home.

I did not grow up in Boston, but it has been an important city in my life. When my father left Brazil this is where he came, and where he met my mother. I lived in the city for several years when I worked in the business world. After I moved to Boston, my mom pointed out that, unknowingly, I had picked an apartment a few blocks from where she had chosen to live three decades earlier, which was itself only a few blocks from where my father had lived in the Back Bay. When I fly to Boston now I can often smell the sea as soon as I step out of the airport. Sometimes I smell the city even before I step off the plane, especially in winter, when the snow-air mixes with the salt and there can be no question where I am. The smell of Boston is not quite the smell of home, but after 3,000 miles of flight to the city where my parents met, it will do.

The smells of cities are so distinct that it's disconcerting when they occasionally fail to match our memories. Once I landed in New York, in the throes of a summer wave of heat and humidity, the day after a trip from Eastern Asia. I took a cab from the airport, and when I opened the window I felt a gust of the night air, the thick water-air of a sweltering city that would barely cool in the evening. If I'd been blindfolded and had had to guess where I was, I would have said Singapore or Bangkok; somewhere near a warm sea, with a neon-scattering waterfront and outdoor markets thronged with evening diners; a place on which snow might never fall.

More intrepid travelers will also know the experience of flying from a shining steel, glass, and marble airport to the sky-harbor of somewhere smaller or poorer, where there are no jetways, and maybe only a handful of flights per day, and where as the plane parks on the tarmac staff rush toward it. As soon as the door opens you feel a rush of wind bearing new smells, and you know instantly it's a different place; it's special not just because the air is different from yours but because there is no built air, and you walk down the steps—a reminder that we arrive not in a place, but onto it.

If there is a charm to this manner of ending a journey and leaving an aircraft, it's because for many of us it is unusual, and it generally occurs in hot places. Few of us want anything like this experience if we fly home to darkness and sleet. Still, such warm moments are a chance to disassemble the word *touchdown*, to recall old films of arriving royalty or the Beatles disembarking the aircraft named *Jet Clipper Defiance* on even a cold February day; the cover of cloud or the blaze of sunshine as feet reach new ground; a weight of arrival that rests as much on the air as on the earth.

*

I've just landed in Tripoli. We're not staying overnight here; such a trip, in an overlapping of the terminology of airline crews and Tolkien, is known as a *there-and-back*. We've parked the plane, the passengers have disembarked, the cleaners have boarded. We arrived early—helped by a tailwind—and so we have some free time before we must begin the preparations for our return.

I wander into the terminal. It's true that airports are increasingly homogeneous, globalized places, but anyone who thinks that this process is complete might compare Tripoli's airport to, say, Pittsburgh's. I walk past the Libyan families and the Western oil workers, looking forward, perhaps, to their first beer in months after takeoff. I head to the roughly decorated cafeteria, to buy a snack I've come to like here: a tasty creation something like a spinach turnover. I browse in the small shop that stocks shelves of books written by Muammar Gaddafi and a handful of postcards of glamorous old Tripoli, the palm trees on the avenues faded and the address side discolored and a little damp.

Eventually I return to the aircraft and walk to the back of the plane, to where rough metal stairs—*air stairs*, naturally—are positioned by an open door. I sit on them in the shadow of the tail, watching the occasional jet land, from airlines and cities whose names are unfamiliar to me. It's hot, and since passengers can't see me from the terminal, I take off my tie. I eat my Libyan turnover and then a sandwich I made in London this morning.

Airport tarmacs have their own smells, of course, but here is also a telltale hazy breeze, one that mixes the heat and the nearby ocean with the golden dust that accumulates on everything and that I will have to brush from my trousers when I stand up. Soon enough it's time to leave Libya, to fly back up into the common air, to cross the Mediterranean, and Corsica and the Alps and Paris, and then to descend to England's sky.

We bank over Tower Bridge. Not very long afterward I walk under the sky where I have flown, under the lights, timed like clockwork, of the next hour's planes, to meet friends at a restaurant on the South Bank. They ask me how my day has been. Good, I say. It was good. Anywhere interesting? they ask, though they mean it half as a joke; nowhere I might answer would surprise them anymore. During the meal my attention drifts occasionally. How could it be, I ask myself, that I have gone to Africa today and returned? I blink and look around at my friends and the crowded restaurant, at the twinkling glasses and the dark woodwork. And I remember, as if from a dream, the blue sea of air over the Mediterranean, the blaze of an ordinary afternoon in Tripoli, and my lunch on the air stairs, in the shadow the plane brought to Libya and then took away.

Geography is a means of dividing the world—of drawing the lines of political entities or per-capita income or precipitation that best illuminate the surface of our spherical home and the often jarringly physical characteristics of our civilization on it. Aviation both writes its own geographies and reflects older ones, as does every air worker and traveler.

There are places I have flown to, and places I have not. This is a way of thinking about the planet that I had not anticipated before I became a pilot; it is one that arguably matters more, not less, the more you travel. On a long-haul pilot's own map of the world some cities glow with frequent and recent experience, others less so, and some are entirely dark. As a relatively junior pilot, I have a map sparser than those of most of my colleagues. It still happens once or twice a year that I fly to an airport I have never flown to before, because the route is new, or the airport itself is new, or the route has switched to the 747 from another aircraft. For days in advance of such a flight I will look at the charts for the airport and for others nearby, or at the flight documents prepared for a previous day's flight. It is common, when we meet our colleagues for a flight, for the captain to ask: Have you been there recently? Or: Have you been there before? We are sharing our maps.

Aside from my personal borders between the places I have and have not been to, the most fundamental division of the world may not be an obvious one, such as whether you are over land or water, in cloud or in the clear, whether it is dark or light. Perhaps the simplest bifurcation of the heavens is between the regions of the world that are covered by radar and those that are not. On the ground at certain airports, markings on our charts exactingly delineate those aprons or taxiways that cannot be directly seen by the controllers in the tower. The whole world is divided in a similar way, by the presence or absence of radar coverage. A surprisingly large portion of the world has no civilian radar. There is none over the seas once you leave coastlines far behind. There is none over Greenland, large parts of Africa, or significant portions of Canada and Australia. Where I fly within a certain distance of a radar site or installation—*radar head* is the term sometimes used for the rotating part—the air-traffic controllers can "watch" my plane in a direct sense. Where there is no radar, they cannot, and we must report our positions via various increasingly sophisticated electronic means or by reading out to them on the radio our time and altitude for various locations, a *position report* that they must then read carefully back, to check that they have heard us correctly.

This sense of being watched, or not, divides the world. To be outside radar range is not like being in a place without cell phone coverage, because we are still in communication with the controllers. It's not like entering a tunnel in a car and losing your GPS location, because pilots know where they are. Nor is the difference comparable to situations in which you are made uncomfortable by being observed, because pilots prefer controllers to be watching them; there is relief when controllers tell us we are *radar identified*, and the sense that we are crossing into a less isolated portion of the journey, or nearing its end.

Mountains above a certain height constitute another division of the world, a separate realm of sky. The altitude above which we are required to wear oxygen masks if the cabin pressure fails is 10,000 feet, and so this rough contour shaped by peaks and an added safety margin forms perhaps the map of the world that a pilot might draw most easily from memory, as if sea level had risen by about 2 miles. The world that remains exists largely in two great, distinct bands. An enormously long swathe of Eurasia, from Spain across to the Alps and the Balkans, from Turkey roughly eastward to China and Japan, crossing the highlands of such countries as Iran, Afghanistan, India, and Mongolia, forms the heart of the map. Another long line of minimum altitudes marked on our charts in red runs in an all but unbroken line along the western side of the Americas, from Alaska down through the Andes; from the Arctic to the Southern Ocean.

On this map of world-height, the United States east of the Mississippi does not exist. Huge portions of Africa, Brazil, Russia, and Canada are absent, too, as is the entire continent of Australia. A similar but inverted sort of blankness covers the peaks of the Himalayas. In 1933, only three decades after the first flight at Kitty

Hawk, the peak of Everest was overflown by an airplane, though one of the onboard photographers passed out from a lack of oxygen. Today there are few routes over much of the Himalayas—not because airliners cannot easily overfly even Everest, but because the terrain beneath limits their ability to descend in the event of technical problems. For this reason many pilots consider the planet's highest mountains least often of all.

The air of the world is divided in other ways. We cannot fly just anywhere. Large regions of airspace are restricted, often for military use, while many smaller chunks are blocked off because they lie over noise-sensitive areas—the center of a city or the palace of a sultan. These restricted airspace blocks are usually marked on our charts by combinations of letters and numbers, not names. But near Mumbai is one known as the Tower of Silence. In the city is a structure on which members of the Parsi community can ritually leave the bodies of the deceased to be consumed by vultures, a process that is elsewhere called a *sky burial*. The area and its name are marked in red on our charts. Some areas where no jets will fly have a ceiling and stop at a certain height, but the Tower of Silence goes all the way up.

There are, of course, great socioeconomic divisions in the world that airliners cross almost as if they do not exist. Even poor countries generally have internationally standardized air-traffic rules and control services. We can envisage in the sky a kind of continuous space, an insulated sphere above and around the earth, in which these standards prevail, regardless of the conditions on the ground below. A plane flies through this well-regulated realm, over cities and countries where we would not wish to land if we had an ill passenger onboard, places that in terms of certain medical services might as well be the ocean; and then we descend from this upper world through similarly regulated corridors down to our destination itself, where a long list of standards—from the suitability of the available water to various safety-related aviation functions—have been assessed. An airliner bound for certain cities will leave London with water onboard for both the outbound and return journeys—sometimes carrying even round-trip fuel and food as well.

In Cape Town, if the wind is from the north, you land from the south, flying, in the last minutes of the flight, near Khayelitsha, a Xhosa name that is almost as beautiful in English—New Home— and Mitchell's Plain, townships that each house hundreds of thousands of people. When I have flown there as a passenger, and had the time to look, I have been struck visually by the power of birth and circumstance: the picture of inequity made by the shining wings from somewhere far away crossing over these settlements, and the freedom international travelers have to descend over the morning of half a million people, some of whom will have already flown, or will one day fly, but many of whom probably never will.

It occurs to me, when I am flying over Hokkaido or rural Austria or Oklahoma City, to ask who might look up and see the contrails of the plane light up at dawn. I feel this equally when I am on the ground looking up at a plane, on the other side of this greeting, as if I'm still a kid marveling at what it must be like to be way up there or remembering my first flight. But there are many places where such reciprocity cannot yet be reflected, the places the plane moves freely over, where place is as heavy as lead.

The most curious aspect of the pilot's life may not be that we work in the air. It is that our world on the ground—the realm of places we know well, and that we connect to other places, the world that for a child begins with the rooms at home, and then expands to the backyard, and to the neighborhood—is so enormous. The job induces an almost planetary sensibility, a mental geography that rounds countries and continents as easily as you follow turns in a path through a familiar wood.

As a child you are taken to places by others. When I grew up and learned to drive I eventually drove myself to many of these same places: small towns, or lakes, or state forests in New England where my family had camped or hiked when I was young. I realized that although I remembered the sites well, they had floated freely in my memory, untethered to actual geography. I hadn't known how they lay on a map or on the earth, how to travel to one, or between two of them, or how long such journeys might take. But when I drove to them myself, the cloud they formed began to sort itself, to fall into place, as we say, like the pieces of a wooden puzzle. I realized that a lake I thought faced in one direction actually faced another, for example, and was close to a second location that I had never linked it to.

When I learned to fly, such a sorting of idea-places onto the physical world around me happened on a fully planetary scale. What suddenly appeared in the window included not only the few cities I had flown to as a child, but everything I saw from the air that was identifiable—all the cities and mountains and oceans I had heard of or read about and dreamed of someday visiting.

This sense of a formal knowledge of places falling onto actual earth and lining and connecting up, one with another, may be similar to the ways in which bodies change in the minds of medical students when they first learn how the organs and bones they've always known the names of are really located in three dimensions, and how they're connected by other tissues they did not know about before medical school. The first time I flew to Athens, I noted some digits on our paperwork that marked the presence of an area of high terrain not far off our route. As we approached this region a snowy peak came into view. I said to the captain: That's quite a mountain. He looked at me as if I had said something strange and then answered: Mark, that's Mount Olympus.

I'm over Arabia, routing toward Europe. Ahead is Aqaba, the lights of the Sinai; then the city of Suez, and the lights of ships streaming through the canal like blood cells in an animation of the earth's circulatory system; then the glow of the Nile, a flowing ring of light around the waters, a flume that fans into Cairo, leading the eye to Alexandria, pooling on the coast; and off to the right is all of Israel, shimmering on its water so marvelously that if I did not know which coastal place I was gazing upon I would bet it was Los Angeles; and beyond is Lebanon, where I look for this night's electrified shadow of the biblical city of Tyre, that "dwells at the entrance to the sea." The next lights are those of ships, and then comes the illuminated net of Crete, and the city of Heraklion. An idea of these places was all I had, until I saw them turning below me in their natural order.

A few hours later I am over Germany, looking down at an inland sea of light. I remember a childhood fascination with an atlas of the world owned by my parents, in which the most densely populated areas were clearly marked out, so that London or Los Angeles or Tokyo were surrounded by splotches of bright red. In northwest Germany, too, there was a large red area like this, that I was certain must be a misprint because it was so enormous and sprawled over a region that was far from any major German city I had heard of, such as Frankfurt or Munich. I asked my dad, and

he told me the name of this place, which I still find beautiful, perhaps because I can remember him pronouncing what I myself can
never properly say: the Ruhr. It is Germany's most populous area,
he told me, though you won't yet know the names of even the
largest cities that constitute it: Dortmund, Essen, Duisburg. The
Ruhr is easy to see from the sky at night, a sprawling illumination
as clear as any colorings in a childhood atlas.

Before I became a pilot I had the naive sense—a feeling, as
opposed to what I'd been taught to the contrary—that most people lived in a world that looked something like where I grew up:
small towns, forests, fields, four seasons, hills in some recurring,
familiar pattern, the reference of a coastline a few hours away, and
the vague gravity of a major metropolis at some similar distance.
Today I understand in a direct and visual sense what I learned in
school, that humanity is concentrated in a dense set of the lower
latitudes of the northern hemisphere, and further in a dense set
of longitudes in the eastern hemisphere; and what I have read
since school, that ours is an age of cities—small ones as well as the
conurbations like Mumbai, Beijing, and São Paulo that dominate
the urban earth—in which for the first time a majority of humanity lives.

A plane's center of gravity is a critical piece of information that
pilots receive before takeoff; it depends on the weight and location
on the aircraft of the passengers, cargo, and fuel. Methodologies
vary, but several calculations place humanity's center of gravity,
the geographical midpoint of the world's population, in or near
the far north of India. I often fly not very far south of there, either
approaching Delhi itself or passing it en route to southeast Asia. I
imagine a bull's-eye of concentric rings that begins at the center of
gravity and echoes out, each ripple eventually sweeping up more

and more of the planet's population, and then I am reminded that I, like nearly everyone I know, am from the provinces, from the periphery of the map when the map is weighted by individual lives. When I fly between New York and London it is easy to forget that only in an economic sense are even these cities much more than outer stars in the galaxy of human geography, and that the place I myself have been centered—rural New England—is almost comically tangential, not even a footnote in the textbook a visitor from another planet might write about the geography of our species. The question of how the world looks to most people is one I would have got entirely wrong before I became a pilot.

A separate question is what the surface of the earth typically looks like. If someone had asked me this before I became a pilot, my answer would inevitably and provincially have focused on what I had seen of the earth in the places where I had lived or traveled—trees, rolling hills, small towns between big cities. Today I would answer that question differently. I would say that the world looks mostly uninhabited.

Most of the earth's surface is water, of course, and an enormous portion of what isn't water is very sparsely populated—whether because it is too hot, too cold, too dry, or too high. We forget this, if we ever even learned it, because we never see it—unless, of course, we look out the window of airliners, at the vast, nearly empty regions that planes bear witness to, and carry us over, the in-between places that are such obvious features of our planet's face but that by definition we are unlikely otherwise to experience. By one estimate, the portion of the earth's surface on which an unclothed human could survive for twenty-four hours is about 15 percent. That's a hard calculation to make—it depends on the season and weather, for example—but from the cockpit of a long-haul airliner, at least, such a figure does not surprise me.

The shock of a nearly empty world is most startling on routes that take us into the far north, where so much of the planet's emptiest land hides in plain sight. Over Canada and Russia, the world's two largest countries, are many hours of flying where you see almost exclusively snow and ice, or their brief seasonal absence; this is the taiga, the forest, and the tundra where almost no one lives. The entire population of Canada is smaller than that of greater Tokyo, and nearly all Canadians live in a narrow strip along their country's southern border. Siberia alone is larger than the entire United States, larger even than Canada; but Siberia has fewer inhabitants than Spain. Northeast Greenland has an area comparable to the combined size of Japan and France, and a population of forty. Many hot places, too, appear similarly desolate. We forget, unless we cross it as often as long-haul pilots do, that the Sahara isn't much smaller than the United States; then there are the vast, barely inhabited portions of Australia, a continent comparable in breadth to the contiguous United States (as Australian postcards that overlay maps of the two make so clear); and then there is the Kalahari, and Arabia.

I don't mean to suggest that the portions of the earth that look empty have not been disturbed—nearly all of them have been, not least by climate change, to which the planes that carry us over such places make a growing contribution—or that we can make useful assessments of our impact on the environment from casual aerial observations. Only a specialist can look down on a brown autumn landscape of Canada or Finland, for example, and say where the snow would likely have fallen by this date a hundred years ago.

But if you have ever hiked or driven through a very rural area or
a nature reserve, and looked closely at the many lesser peaks that
surround one well-known mountain, and speculated on whether
anyone has ever stood on them, or even whether some have ever
been given a name, then that is exactly the feeling I often have
while looking out from the window seat of a long-haul airliner. In
all contradiction to what we know about our negative influences
on the world, so often from above it's disturbingly easy to imagine
that we are the first to look upon the earth, that we are seeking a
level place to set down our ship.

The author J. G. Ballard wrote that "civility and polity were
designed into Eden-Olympia, in the same way that mathematics, aesthetics and an entire geopolitical worldview were designed
into the Parthenon and the Boeing 747." Those who fly often may
naturally acquire the worldview, inaugurated by the 747, that takes
a planet to be a reasonably sized thing.

I've come to measure out countries in jet time. Algeria surprised
me, when I first started to fly across Africa. North to south it is
nearly a two-hour country and I now feel what I did not then know,
that it is the largest country in Africa. Norway, too, was another
surprise, on routes to Japan that give us this country from end to
Norwegian end; in the north of a continent crowded with smallish
countries it is a fully two-hour land. France at the angles I most
often cross it is a land of around one hour, as are the states of Texas
and Montana. Belgium, with a healthy tailwind, is a fifteen-minute
country. On many routes Russia is a seven-hour country, though
really it's best imagined as a day-long or night-long land.

I often fly over tiny, windswept Heligoland, an island in the German Bight of the North Sea; there is an important beacon there

that many pilots will know. Britain once swapped its Heligoland for Germany's Zanzibar, off East Africa. Pilots may swap cities, countries, continents just as nonchalantly. I might give a colleague a Johannesburg on Monday for their Los Angeles on Tuesday, or exchange a Lagos for a Kuwait. Some crew find their body clocks prefer one time direction over another, and so they will say that they "do better east" or "do better west" and may ask to swap with a colleague of an opposite-pointing disposition. I generally prefer west to east, though I'm still occasionally surprised to hear myself talking about cardinal directions as if they were brands of breakfast cereal.

I might eat dinner with a member of the cabin crew at a Belgian restaurant in Beijing, and he may ask if I know this or that Thai restaurant in San Francisco, or a new café in Johannesburg that he heard about on his last trip to Sydney. Countries blur, cities elide. Airline crews experience this age of cities, if not quite as casually as they do the rooms of their homes, as little more than different districts of the earth metropolis. Someone asks me if I can recommend a good spot for breakfast in Shanghai. I pause— I cannot think of a place. It takes me a moment to remember that I have never been to Shanghai.

In contrast with the various species of lag and the occasional loneliness of my job, I have enjoyed connections that would otherwise not have been possible. There are many friends from high school and college whom I see regularly only because I am a pilot and can fly to the far-off places in which they now live. When I flew short flights within Europe, I was able to visit relatives in Belgium and Sweden almost as often as I pleased. It was as if a switch had been turned on, and entire branches of the family tree were re-illuminated. I would move so often and so casually around

Stockholm that my once-unremarkable inability to speak Swedish began to seem increasingly strange to me.

For many years, too, I would go to Paris at least once a month. One afternoon I went to the Musée Rodin, and as I walked down the rue de Varenne I had a sudden memory that this was the neighborhood in which my mother had lived when she studied for a year in Paris. She had told me the name of this very street, perhaps, as part of her story about the French strangers who gathered around her to offer condolences, when they heard her accent on the day that President Kennedy was shot. I called her in Massachusetts, interrupting her breakfast. Yes, she said, with a smile I could hear even as her voice drifted off, it seemed, into memory; she had indeed lived on the rue de Varenne. I took many photographs of the street and her old building from different viewpoints, and mailed them to her.

I had a pen pal in Australia when I was a child, a friendship that was held together by the tissue of aerograms until the night, two and a half decades after we had first written, when I myself was charged with flying the airmail to Australia. My colleagues and I flew from Singapore across Indonesia, then across the entire outback to Sydney, where after a long sleep and an enormous cup of coffee I met my pen pal for the first time ever, at a bar on the promenade below the Opera House. Australia will never feel close to me. But that one month it had nevertheless appeared on my schedule, in the small-print, all-capitals form of the ordinary airport code, SYD, and the long arc of a childhood transoceanic friendship—formed precisely because there was almost no chance I myself would ever travel such a distance—was closed.

A pilot's awareness that the whole world is possible is most acute when on standby. Sometimes these standby duties take place at an airport, but often they are assigned at home, where we are required to be reachable by telephone, and to be within a certain travel time of the airport. When another pilot cannot make it to work—illness, a child-care problem, a flat tire—then a standby pilot is called. Sometimes by the time I reach the airport the passengers have already boarded, the fuel and cargo are loaded, and staff are standing by the airplane's last open door, waving their hands at an oblique angle that suggests both a greeting and the direction in which I should keep moving, through the airplane door that they will close as soon as I pass through it.

When I'm on standby I have a bag permanently packed, with both winter gloves and a bathing suit amid the uniform shirts, a bag for all lands, all seasons. I'll be at home cleaning, or at the supermarket, or running in the park, when my phone will ring and a voice tells me that I'm bound for Bangkok or Boston or Bangalore. I return home, pick up my bag, and fly there.

Occasionally I think that the more broadly you wish to experience the world, the more certain it is that you would enjoy being a pilot, even if flying itself isn't your first love. Alfred de Musset, in a sonnet dedicated to Victor Hugo, wrote that in this "low world" you should love many things, in order to know at the end the thing you love best.

He lists some things we might love. He includes the sea and the blue of the heavens, and few pilots would disagree. But a pilot might also take the "low world" more literally. If you are interested in motorcycling or urban design, opera or kite surfing, hiking or languages, the whole world of these things opens to you, at least on long-haul trips, where even if you are able to fall asleep for a half day after arrival and before departure, you often have an entirely clear day or two in between. In many cities—Bangkok, Mexico City, Tokyo—cooking schools offer short courses that are popular with pilots and cabin crew, an opportunity to ground ourselves in the tastes of a new place and to do the cooking that a life of hotels and restaurants cuts us off from. Sometimes everyone at such a class works for one airline or another, sharing a few hours and a table in a far place.

Many of the pursuits that fill our off-duty hours are aligned with the natural world. I have flown with pilots who explore botanical gardens wherever they go, or put time changes to good use by rising early to photograph sunrises. If you have an interest in wildlife, you will have the opportunity to see pandas or elephants or tigers or whales where they live, or to see birds in one season and continent, and then later in another; to overfly their migrations and wait for them somewhere else. You might read a book about the great trees of the world and a few years later have stood under many of them in their original habitats.

I don't mean to diminish the many challenges of the job— the initial training costs that burden most new pilots with home mortgage–sized loans; the days and nights and holidays so far away from family and friends; the irregular schedules that make regular commitments to neighbors or local sports teams or community organizations so challenging; the permanent regimen of biannual, multiday simulator exams; the physical stresses of unpatterned night shifts, time-zone changes, and other circadian upheavals; and the knowledge that our livelihood can vanish entirely in the furrowing of a doctor's brow during one of our regular medical checkups. Flying is work in every sense. ("The reason you fly," admonishes the father of Jonathan Livingston Seagull, "is to eat.") But there are few jobs, I think, whose side rewards are so varied—as wide as the whole world—and so freely determined by the individuals involved.

One of my own interests when I am away for work—*downroute*, as we say—is hiking, which seems to help with both place lag and jet lag, although whether this effect is due to the exercise or the simple act of placing my boots on the soil, I do not know.

I'm walking in a park in South Africa; it is hot and dusty; I am here with several of the flight attendants and the pilots. The soil is red and the sky is blue; we are sleepy but the sun and conversation keep us awake. It was nearly freezing when we left London last night; overcast, late autumn, the anti-icers running on all the engines for takeoff. Dawn came over Botswana, and as we descended together toward Johannesburg a few hours ago, we saw this land below, this color, smoothed to perfect abstraction and running to the horizon as the sun rose on this spring morning in southern Africa. Now I'm walking on the land we crossed over. This soil gusts up in small crimson clouds with each step of mine that falls on it. A colleague points to a tree, to a weaver nest hanging from a branch; he tells me that the birds are named for their skill in making these nests.

It's four days later. I'm at home, standing sleepily by the sink. The water runs over the soles of my sneakers, sweeping the African dust brightly over the stainless steel. I have to say it in my head, practically spell it out: "This is the red of the soil under the South African tree, from the morning I saw the weavers and their nests." I think of the term *earth*, both soil and planet; this earth could not have expected to meet this water, here. People become quickly accustomed to peculiar aspects of any job. I try hard to remember that this is an unusual experience of the world—to have stood on the earth there, then there on it and there, then suddenly to find myself alone on an ordinary afternoon, quietly washing it from my shoes.

Wayfinding

In 1904, at a time when pocket watches predominated among men, the great Brazilian aviator Alberto Santos-Dumont asked Louis Cartier to make a wristwatch for him, to help him time airborne events without lifting a hand from the aircraft controls.

Today I am required to wear a watch to work—digital or analog—and to check its accuracy before departure. All aviation runs on a single time zone, variously called UTC (Universal Time Coordinated), GMT (Greenwich Mean Time) or Zulu, the last in an alphabet of military abbreviations for time zones. Schedules are written in it, and it is the only time that the airplane computers know or display. When I make an entry into these flight computers, I distinguish the time 1400 hours from the altitude of 1,400 feet by appending a Z, for Zulu, to the former. My work schedule is printed in GMT each month. When an airport publishes the hours of a temporarily closed taxiway, or the day and time a line of thunderstorms is expected to arrive overhead, all this is written in our global time. Often not only the time but also the date of the crew's departure is different from that of the passengers on the same flight. Your Monday evening departure from San Francisco takes place, for me and everyone else working on the plane, on Tuesday morning.

Language, too, is standardized. Commercial airline pilots, regardless of nationality, will speak English; we also share the technical dialect of aviation. I may never meet a 747 pilot from China or Germany, but if I did I would be able to discuss my work with them, in English. The language of the labels in the cockpit is English. When the plane itself speaks to us out loud, it speaks in English.

Worldwide, air-traffic controllers can speak in English, but a set of "English" terms so specific to aviation that few who are not pilots or controllers would understand the distinctions they contain. In certain countries, controllers speak to local airline pilots in their local language, and such pilots will feel more at home there. But in much of the world's busiest airspace, controllers will assiduously adhere to English. The truth of globalization is never clearer than when, for example, I arrive overhead a German airport and German controllers are speaking to German pilots, but not in German.

If aviation is commonly associated with the leveling of differences, with the bulldozing of borders between places and times and languages, it has also resulted in the creation of new realms of geography—a new world, high above the old one, that is not yet fully charted.

The sky is divided into administrative divisions of airspace. These divisions aren't straightforward; there are various, often overlapping kinds, and often the name of a region on a map is not the same as that used to identify the controllers and radio operators we speak to there. The sky regions may be roughly equal to an earthly place you have heard of, or smaller, or much larger. Those that cover coastal areas may soar out from their littoral realm, across dizzying swathes of open sea, until they meet a region that does the same from another continent, and these two boxes of air meet at the air wall between them; coastlines that meet only other air-coastlines. At the poles many regions meet, the common points of many slices of aerial pie.

Though they are not well matched to terrestrial places, these regions have borders. Their names have histories. They are the countries of the sky.

All Japan lies in one region. The name of it is not Japan, but Fukuoka; yet within this sky country marked on the map we speak to controllers who answer variously to Sapporo Control, to Tokyo Control. America's regions look much as its states might, if some pitiless war or committee had hugely reduced their number. Salt Lake City (abbreviated to Salt Lake, as in "contact now Salt Lake, on frequency 135.775") covers parts of nine states, from southern Nevada, north over the Great Salt Lake itself and its city to the Canadian border, which it meets between the sky states of Seattle and Minneapolis. Southern Illinois is not part of Chicago; it's divided between the dominions of Kansas City, Indianapolis (sometimes called Indy Center), and Memphis. There is a region called New York; yet most of New York State lies in Boston, which also encompasses all of New England.

In contrast to America's consolidated sky states, many small European countries have kept their own regions. Switzerland has done so; its sky land is called Switzerland but it's Swiss or Swiss Radar that the controllers answer to. "Swiss, good evening," I might say, followed by my call sign. The region is so small, and planes so fast, that a jet may cross through it in minutes. Greek controllers may answer to Athens Control but their sky country is marked on charts as Hellas—Greece. On the busy routes along the Adriatic coast, a plane may be in the sky country known as Beograd—Belgrade—for only minutes; the controller there will say hello to us, before almost immediately transferring us to the next shard of the former Yugoslavia.

The sky known as Maastricht makes an auspicious contrast to such fractured air. Maastricht is the legacy and incarnation of a high, early dream of European integration that's now sometimes called the *Single European Sky*. Perhaps the best-known volume of air to most European pilots, Maastricht covers the higher airspace of Belgium, the Netherlands, Luxembourg, northwest Germany, and certain nearby areas, a whole and peaceful dominion that rises over some of the continent's historically bloodiest borders. I've been to Maastricht, on the ground, but if you said the name to me, I would not think of the Dutch city, of earth-Maastricht. I would think instead of sky-Maastricht, this invisible block of the heavens resting on the fragmented history of the northwest corner of the continent. Sky-Maastricht is not Belgium or Luxembourg or the Netherlands, yet its cold aerial polyhedron, sharply bordered and as meaningless as sliced air, blankets them all—a new, improbably named country above Europe.

The names of air regions may not correspond to any place on earth that is familiar to me. The syllables then form a kind of aerial poetry, a drumbeat of far sky-lands beyond the next fold of the chart: Turkmenabat and its sister Turkmenbashi; Vientiane, Wuhan, and Kota Kinabalu; Petropavlovsk-Kamchatsky, Norilsk, and Poliarny. Or the names match those of legend, those that might be among the last you would expect to rise to such prominence in the modern sky: Arkhangelsk and Dushanbe; and Samarkand, the city that Marco Polo and Ibn Battuta wrote of, which fell to Alexander the Great and Genghis Khan.

When I come across such far-sounding names I wish I knew how Maastricht sounds to, say, an Uzbek or Chinese pilot; if the peculiar lettering, the twinned *a*'s and the *cht* contradict years of English lessons, to form a sound so unusual that it can only be from a curious place, the inland city of a small and venerable sea-faring land. Or if a Dutch place name, heard or imagined from the other side of the world, instead sounds hardly different from English, a linguistic truth that such pilots see borne out from the sky, where the Netherlands and southeast England may appear to rest only a few grammatical rules or major drainage projects away from each other. The question of what distant place names sound like arises in its purest form in the sky, where the pilots from all places cross all places.

Other region names are familiar to pilots before we fly anywhere. London, Delhi, Bangkok; world cities beneath their eponymous air countries. Flying into these regions, it is almost as though we are entering the city's aerial sphere of influence, the gravitational reference of the metropolis, as if we're caught in its conceptual— and at night often its actual—glow. The grandeur inherent to the name of an enormous city is heightened further when American controllers occasionally append *Center* to the name of a region, preceded perhaps by *The*, as if part of a monarch's style, or a city's formal epithet. "Contact now The New York Center" has a certain ring to it, if at the moment we receive such an order the night is filled with the cloud of light rising up from the city itself.

A sky country off West Africa is called Roberts. When I first saw this I was reminded of Robert FitzRoy, the meteorologist and the captain of Darwin's ship HMS *Beagle*, because one of the areas in the BBC's Shipping Forecast, FitzRoy, is named for him. These areas are an analogous kind of sea country, the white-capped conditions of which we may observe from above, and the names of which I became accustomed to when I flew early departures on the Airbus and would leave home long before dawn, drinking coffee and listening to the radio as I drove to Heathrow through the nearly empty wet-black streets of London. Roberts, the aerial region, takes its name from Liberia's first president, Joseph Jenkins Roberts, who was born in America and moved to Liberia when he was twenty. An enormous African sky will bear his name forever.

In Britain, London is a noun; yet the region north of it is an adjective, Scottish. "Contact now Scottish," a London controller will say in farewell to a northbound aircraft. To a southbound jet, meanwhile, a Scottish controller will say: "Call London," and it is hard not to think of the BBC World Service identification, "This is London," or "This is London calling," the old voices that shared this air.

Many regions have grand, waterborn names. Above South America lies Amazonica. I like it when a controller tells me to "contact now Rhein," if I can see the river below. Many regions encompass vast swaths of airspace above oceans and reflect this in their names. There is Atlantico, which fences off a large sky in the central and southern Atlantic. The names Anchorage Oceanic and Anchorage Arctic are stretched over stormy, gray-and-white seas. They might be the names of ships. An enormous portion of the open Pacific is held on maps in the air-name of Oakland Oceanic, though pilots will speak to radio operators who answer to the name of San Francisco—a cross-bay rivalry drawn out over much of the Pacific. It is Oakland, though, that is the name of the sky, an ocean-straddling aerial empire whose extent might surprise the city's residents, as might their city-sky's borders with Manila, Ujung Pandang, Auckland, and Tahiti.

There is air over northern Cyprus that is claimed by two regions, and so we speak to two controllers, on two different radios. There is a sliver of airspace off Norway that does not lie in any sky country at all. This no-man's-sky splits Norway's Bodø and Russia's Murmansk like a knife, as if it were created by some blistering of the skies since they were first charted, or in an aerial version of how new islands rise from volcanoes in the sea. Another remaining nameless realm of sky lies in the Pacific, west of the Galapagos, north of the sky land Isla da Pascua—Easter Island. These blank spots are not what we would expect to find in the realm of the airplanes more often associated with the dispelling of the world's final mysteries of place.

In Africa the region Brazzaville answers to Brazza. The quality of radio transmissions is not always good here and it is often said twice, loudly. If you say to long-haul pilots, in a clear strong voice: "Brazza, Brazza!" they may smile and think back to the early hours of nights that passed under equatorial African stars. Two sororal skies of West Africa are perhaps the world's most gracefully named: Dakar Terrestre, which beyond the coast falls away to Dakar Oceanique. Here is Dakar, its earth-sky and its sea-sky.

There is a majesty to the borderlands, where pilots will transfer from one set of controllers to another. Here a flight, as we say, will be *handed over*, from one place-name to the next, and so on over the world. Often one region will give us to the dominion of another a few miles before the actual border. "Call now Jeddah," the controller will say to us, "you are released."

*

A pilot may acquire an affectionate awareness of a kind of punctuation or asterisking of the world, composed of the names of
small places, places that almost no one other than pilots will have
reason to think of regularly.

Many milestones are elevated this way because they are home
to a radio beacon. It's hard not to think of older beacons, lit to
help navigate, as with lighthouses, or to transmit warnings, as with
the news of the sightings of the Spanish Armada, or to celebrate
events such as coronations and jubilees. In the 1920s, hundreds
of light beacons, often placed high on mountaintops, inaugurated
the first transcontinental airmail flights, from New York to San
Francisco. This cross-country trail of light echoed the railroads,
of course, but also the Pony Express, as pilots and planes would
change en route, allowing letters to make an almost continuous
journey from one coast to the other. There was even talk of a
"lighted airway" across the Atlantic for dirigibles. Today, particularly in the western United States, some of the radio beacons used
by modern airliners are sited just where those original light beacons once stood.

Pilots can manually tune a beacon and see our distance and
bearing from it, a basic, old-school check of our position. But in
the background a modern aircraft is always searching for them,
like a driver in an unfamiliar town constantly seeking landmarks
and street signs. Beacons have only a certain range, and when
the aircraft finds one, its codes may flicker to life on one of our
screens, and in this way we come to know the names of many of
the beacons of the world.

Near the tip of Cape Cod, on the ocean side, stands a beacon
that is a curiosity to those walking near the beach and that shares

a name, appropriately, with Marconi, the Italian engineer known as the father of radio. Beacons like this one and the plane speak to each other, like children playing Marco Polo in a pool. The plane counts the time between its call of "Marco!" and the beacon's reply of "Polo!" and so calculates the distance between them.

In the more remote regions of the world beacons and airports often coincide; the beacon is there because the airport is there. When such a place is surrounded by nothing else that relates to aviation, its isolation lifts it into unexpected prominence in the sky. In Greenland is the airport named Aasiaat. It is on a bay I would like to visit one day, because it bears the marvelous name familiar to armchair atlas ponderers, long-haul pilots, and almost no one else: Disko Bay. The names of many small places in the far north of Canada have the quality of making bitterly cold water sound warmer than it can ever be—Pond Inlet, Sandy Bay, Hall Beach, and Coral Harbor. There are airports such as Churchill's, in Canada, that are the only suitable runway for many miles in any direction. Often as white as paper, Churchill is habitually visited by polar bears; it stands on Hudson's Bay, where Hudson and his son were forced off their boat in a mutiny, after the ship was freed from the ice that had immobilized it through the long winter.

Also in Canada, listed on our charts is the place called Gjoa Haven, named by Amundsen for his ship, the *Gjøa*. Amundsen was there to look for the north magnetic pole. In general terms, the closer you get to the magnetic pole, the crazier an ordinary compass becomes, as if you were approaching some fearful, caged creature. Gjoa Haven appears on our maps near the dotted lines that formally designate the Compass Unreliable Area, which is near the Compass Useless Area, further unexpected divisions of the modern sky and the world.

Some beacons are in places that although famous are geographically incidental; you might not expect them to be elevated on aviation charts in a manner so independent of their historical prominence. Point Reyes is the name of a lighthouse on the Northern California coast; a beacon near it, known by the same name, features on arrivals in San Francisco. On flights over India, we may fly over the beacon of Delhi, and like so many Taj Mahal–bound travelers below, our next stop is Agra. Robben Island, off Cape Town, where Nelson Mandela was incarcerated, was a prison even in the seventeenth century. It's home, too, to a beacon of the same name, which appears on charts for Cape Town's airport, and forms part of an often-used arrival pattern.

I have a Canadian friend from a small town in interior British Columbia. When I first asked where she was from, she laughed and shook her head and said I would not know it; it was a tiny town where they didn't close the school unless the temperature was colder than minus 40 degrees Fahrenheit. But when she said the name of this small place, Williams Lake, it was my turn to smile and say: I know Williams Lake; I gaze on it every few months. There's a navigation beacon there. When I see her, if I have flown over it recently, I will tell her if it was cloudy, or if I could see her hometown resting between the Rockies and the Coast Range.

In Japan's Ibaraki Prefecture is a place called Daigo, a town of some 20,000; foreign pilots will probably not know about its waterfall, but they may know its beacon. Hehlingen is the name of a village in Germany; it is also the name of a beacon sitting in a nearby field, a name that bounces more or less constantly around the skies between Hanover and Berlin, in a range of accents as wide as the world. There are beacons named Split Crow, near Halifax in Nova Scotia; and Old Crow in the Yukon.

There is the Rome of southeastern Oregon; and Norway House, which is many flying hours from all the dwellings in Norway. There is Muddy Mountain and Uranium City; Crazy Woman and Vulcan.

The names of other beacons are more mellifluous. In Scotland I occasionally overfly Machrihanish, a coastal village from where a message was sent to my home state of Massachusetts in the early days of radio; the beacon appears in the cockpit as MAC, spelled out as "Mike Alpha Charlie" by pilots like me, who dare not try the full name in a conversation with a Scottish-accented controller.

In northern China, set in the tawny elevations of the Gobi Desert not far from the border with Mongolia or the railway that connects the two countries, is the beacon named Eren. In north-central Pakistan, on the west bank of the River Indus, is the city and beacon named Dera Ismail Khan. In Algeria is Bordj Omar Driss, bearing the identifier BOD, which a controller will pronounce as "Bravo Oscar Delta" to a pilot who does not know this small Algerian town, population around 6,000, by its actual name. Russia has many fine beacon names: I like Maksimkin Yar and Novy Vasyugan; my favorite is Naryan-Mar, a coastal town of some 20,000 inhabitants, and a welcome milepost beyond the Arctic Circle.

An airplane navigates through the sky along a route composed of beacons and waypoints. Waypoints are defined by geographic coordinates or their bearing and distance from a beacon, and by a name, which typically takes the form of a five-letter capitalized word—EVUKI, JETSA, SABER. The idea is that they will be pronounceable and distinct to controllers and pilots regardless of their first language. The pilot's map of the world, and the flight computers' too, is atomized into these waypoints. They are the smallest nuggets of aerial geography, and in some sense the only such unit that matters once you leave the runway. They are the sky's audible currency of place.

From a plane, even a wide modern road can look as slow and old-fashioned as an ancient bridleway. The plane slides like an eye over the page, like a finger across a map, over everything the road and the drivers on it must turn to avoid—towns, mountains, lakes—features so low they appear nearly smooth from above. Waypoints, though invisible, remind us that while pilots are not nearly as constrained by the sky as drivers are by roads, neither is our path always as free as it appears.

That is not to say that a waypoint is a place like any other. Though they are often strung together in airways, we're frequently allowed to move between two distant waypoints without overflying those that lie between; as if a driver could leave the road to tunnel directly through hills and forests before meeting it again, further along. And a waypoint, for all its extraordinary specificity, is not a single place at all. It exists at all altitudes at once. It is possible for many planes to cross the same waypoint at the same time, at different altitudes, yet each plane's navigation computers show it at the same position. A waypoint is like the address of a skyscraper that does not specify the floor. The speed of a cruising airplane also means that we often do not get anywhere near a waypoint that is on our flight plan, because we must turn well before the waypoint if we are not to overshoot the route on the other side of it. For a sharp turn, in a strong tailwind, we may begin to turn 5 miles before the waypoint, something to imagine, that in a car you would start to turn the wheel so far before the intersection.

There is a rhythm to waypoints, which roughly matches the

rhythm of the human geography below. Tourists from North America wandering the cities of Western Europe may have the sense that historically significant places occur every few dozen yards; in the sky over Europe we may cross a waypoint every minute. In contrast, over open sea, or a place such as northern Canada, we may fly forty-five minutes or more, hundreds and hundreds of miles, between waypoints. The pace of passing waypoints also roughly echoes the workload in the cockpit. Most of the waypoints crossed will come in the first and last minutes of a flight, when the plane must make many turns to move between a runway and its route, and then back again at the far end.

Pilots come to know many individual named points on the routes they fly most often. Some, such as those that are well-known entry and exit points for Atlantic Ocean crossings, feel like doors, almost, or gates—when I think of LIMRI or MALOT, off Ireland, I think of the phase of flight in which they occur, the start or the end of an oceanic crossing. The feeling is comparable to the name of a bridge that you only cross when leaving or entering a city, one to which newscasters will casually refer when talking about traffic, and you know they are speaking to those who are leaving town or planning their return.

The names of many waypoints are random; an example of that early lesson taught in linguistics that there are many more possible words—spellable, pronounceable—than there are actual words. There is an automated tool available to airspace planners that generates just such names and helps ensure that identical names are not geographically close. Many other names, however, are not random. In these we see perhaps the last realm on earth in which meaningful place names are scattered over a geography that is new to the namers, a world that is new, in this case, to everyone.

Many names in the new geography of the sky reflect aviation's
nautical heritage and the water below them. Near Perth, Australia,
are the waypoints FLEET, ANCOR, BRIGG, SAILS, KEELS,
WAVES. South of Newfoundland, in the vicinity of the historic
Grand Banks fishing grounds, is the waypoint BANCS; further
north along the Canadian coast lie SCROD and PRAWN. Sometimes there are multiple waypoints with the same name, and when
we type one into a flight computer, it will ask us which of these
homonymous, far-scattered places we mean to navigate toward.
There are five SHARK waypoints—one east of Sydney, the others off the islands of Jersey, Maui, Taiwan, and Trinidad.

Near the Isle of Man is KELLY, in reference to an old music-hall
song called "Kelly from the Isle of Man." Off England's Channel coast are DRAKE—for Sir Francis—and HARDY—for Sir
Thomas, the old friend to whom Lord Nelson, as he lay dying
on the deck of his flagship, was heard to say: "Kiss me, Hardy,"
and "God bless you, Hardy." On sky maps of the Tasman Sea,
the triangles that denote the waypoints hanging like notes on a
musical staff arcing toward New Zealand are marked WALTZ,
INGMA, and TILDA—a reference to Australia's unofficial
anthem, "Waltzing Matilda"—while many thousands of miles
west, running north to south over hundreds of miles of Indian
Ocean off Western Australia, is a lyrical sequence that begins
WONSA, JOLLY, SWAGY, CAMBS, BUIYA, BYLLA, and
BONGS—"Once a jolly swagman camped by a billabong . . ."

Continental Europe has fewer locally themed waypoints, or
at least fewer that are apparent to an English speaker, though
off the Dutch coast floats TULIP, and it's easy to speculate
about SASKI—Rembrandt's wife was Saskia. Over Germany,
an English speaker might hear ROTEN as a meaningless, albeit

pronounceable word; a German pilot might hear the bells of the medieval town of Rothenburg ob der Tauber. Crossing the border between Austria and Germany are a series of waypoint names that form awkward phrases. NIGEB—DENED—IRBIR is a loose variation on the German *Nie gebt denen ihr Bier:* "Don't ever give them [the pilots?] their beer." In the heavens near Stuttgart are VATER and UNSER, "Our Father" ("who art in heaven," as the Lord's Prayer continues). Northeast of Nuremberg, near the German–Czech border, are ARMUT, "poverty," and VEMUT, *Wehmut,* German's fine old word for "wistfulness."

Near the border of India and Pakistan is the waypoint TIGER. Another TIGER forms part of an arrival pattern for London, as if lifted from Britain's former empire as incongruously as an animal taken from a warm place to a zoo in a cold city. On flights from Singapore to London I may overfly both TIGERs in the same night.

America's sky-mappers have gone to more trouble than most to ensure that local colors fly in the country's skies. The Sonoma County airport in California is named after Charles M. Schulz; nearby is the waypoint SNUPY. Near Kansas City are the culinary waypoints BARBQ, SPICY, SMOKE, RIBBS, and BRSKT. Near Detroit is PISTN, surely for the basketball team whose name reflects the city's heritage of industry; the skies around Detroit also feature MOTWN and WONDR (Stevie, Michigan-born) and EMINN, perhaps for the rap star. Houston's nearby SSLAM is followed a few miles beyond by DUUNK (not to be confused with DUNKK, near Boston, a reference perhaps to a certain Massachusetts-born doughnut chain). The skies around Houston also feature ROKIT for the city's space legacy, and TQELA, WORUM, CRVZA (beer), CARNE (meat), and

QUESO (cheese) for the city's cross-border culinary traditions that arriving passengers may soon be enjoying.

Boston has etched a particularly intricate constellation of itself onto the ether above New England. There is PLGRM, for the region's history; CHWDH, LBSTA, and CLAWW for its food; GLOWB and HRALD cover the city's newspapers; while SSOXS, FENWY, BAWLL, STRKK, and OUTTT chronicle the anguishes of the city's baseball team across the heavens. Even the region's speech—WIKID, followed by PAHTI—seems to be mapped. There's a NIMOY waypoint; Leonard was born in Boston. LYHTT floats above the harbor island on which stands Boston Light, the 1783 replacement of the 1716 beacon that a twelve-year-old Benjamin Franklin memorialized in a ballad. Passengers may cross the LYHTT waypoint and see this lighthouse, the first in what would become the United States and the only one that retains a lighthouse keeper, as they descend to the city it marks.

St. Louis has the nearby waypoints ANNII and LENXX, for reasons that aviation authorities could not explain to me; perhaps it's only that an air-traffic controller there was a Eurythmics fan. The origins of other waypoints near St. Louis—AARCH, for example, a reference to the city's skyscraping Gateway Arch—are less obscure. Mark Twain died seven years after the first flight at Kitty Hawk. The riverboat pilot himself never flew. But *Tom Sawyer Abroad* features a "noble big balloon" equipped with "wings and fans and all sorts of things," and in an 1869 letter Twain wrote that "the grand problem of aerial navigation" is "a subject that is bound to stir the pulses of any man"—reasons enough to think he might be pleased by the thought of the sky place TWAIN, above Hannibal, his childhood home on the Mississippi.

*

As a sequence of places—beacons, waypoints—a route becomes its own kind of place. In the Japanese language there are many *counter* words to enumerate conceptually similar kinds of objects. English, as instructors may point out to students of Japanese, has a few such analogous words. We say three loaves of bread or two sheets of paper, so *loaf* and *sheet* are our counter words for bread and paper. We think of *sheet* as a concept of form and apply it to many flat things such as aluminum and pastry. One's favorite counter is a regular topic of conversation among foreigners studying Japanese. I've always liked 本, *hon*—perhaps because it is also the *hon* in 日本, Nihon or Nippon, Japan, the destination of my high-school summer homestay and some of my most memorable subsequent journeys, first as a consultant and later as a pilot. As a counter, 本 is used for long, cylindrical objects such as pencils, films, roads, and rivers. 本 can be used to enumerate air routes. Contrails, too.

The exact route a plane flies between two cities often changes. Airways are pre-published lists of waypoints and navigation beacons, crossing steadily over the undisturbed and uncharted fields and forests and rivers below. Flight planners and pilots choose among the various airways that can link two cities, accounting for the wind, airspace closures, congestion, and navigation charges imposed by overflown countries. One path intersects another, and a flight plan will often jump between airways at these junctions, tacking a wind-optimized path over the earth. Other routes, like those over the North Atlantic, are drawn anew each day by air-traffic authorities, to reap the most of tailwinds or the least of headwinds. In some parts of the world there are few or no formal airways, nor wind-sculpted schedules of daily routes. In such open sky country flight planners are free to compose their own route each day, from raw points of latitude and longitude—digital fictions in the ether that the broad sunlit wings of a high 747 will overfly long after the planner has returned home, eaten dinner, turned the lights off, and gone to bed.

Often the technical precision of routes barely conceals their historical and cultural resonance. The daily routes over the North Atlantic are typically called *the tracks*, a general term for the thick, highly trafficked belt of routes between Europe and North America, and a daily refrain to the bonds—of exploration, empire, language, trade, culture—that run as deeply as ever across these waters. Over Africa, the predominant flow of traffic is traditionally from north to south, but during Islam's Hajj pilgrimage, a huge flow of east–west air traffic briefly arises, crossing northern Africa toward Mecca, a season of the air that is an echo and a reversed image of the historical flow of Islam itself. There are special Hajj charts and procedures, issued each year to pilots who will be crossing North Africa at that time.

Not long ago we switched from mostly paper charts to mostly electronic ones, stored on tablet computers. The paper charts we used until recently, though falling out of technological favor, are more interesting, because their designers could not place everything on them, nor could pilots simply select different layers of information to display or hide. The choices required by the mapmaker reveal much about the geography of the sky. The paper charts show airports, but not the cities they correspond to. They don't show roads, or the earthbound borders of provinces or states. Mountains are unnamed and exist not as peaks or contours on these charts but as generalized heights for an area. Even the names and borders of countries are not displayed prominently. The most obvious features on these paper maps are complex networks of dark lines, the routes that link waypoints, the highways of the aerial world.

Even the cut of the paper follows routes, which follow history. Most ordinary maps—those in a typical atlas, for example—appear to be rectangles lifted directly from the earth. The top of the map is roughly north, the bottom south. But the rectangular en-route paper charts are often not aligned this way. In their off-kilter calibrations we see the deep axes of empires, migration, and the whole of human geography, as clearly as the archaeologists who made some of the earliest uses of aircraft, to discern man-made patterns on the earth. For example, the paper maps that a pilot may follow from Europe to Hong Kong are not oriented to north. Instead they are cut in long, great-circling arcs that roughly echo the typical routes between Europe and China, so that the left edge of the rectangular sheet faces northwest, and the right edge faces southeast. Over central and northern Canada a series of charts runs in a similar but more sharply tilted orientation that only makes sense in the context of the air links between eastern North America and East Asia.

Routes often have a personal weight. Frequent travelers over the North Atlantic, for example, will have an intuitive sense of just what the plane connects; the cultural and historical bridge that each of their own journeys reflects and renews. The first time I flew across Australia, from Singapore to Sydney, we made landfall on the country's northwest coast, near Broome. Then we followed a series of long airways that arced across the continent toward the southeast. The distant lights of Alice Springs reminded me that I had first read this name in a Hardy Boys story that hinged partly on the revelation that this overheard name was a reference to a place and not a person.

As we crossed the outback it was hard not to think of song lines—of not just the route in the computer but of how everyone onboard might imagine the line of space between the paired cities, our individual assemblies of mental and physical geographies. For passengers a route exists in the moving map on cabin screens, but also in their understanding of how two particular cities differ and why they are flying from one to the other, and of what stands or floats between them. Most of the Australians onboard that night would surely have had a deeper sense of the route than any foreign pilot flying over it for the first time.

Eventually, routes become familiar to pilots, and then the passage of landscapes, sky regions, or beacons give journeys between two cities a unique pace. Pilots will acquire this background sense of the progression of a route just as the conclusion of one song on a beloved playlist leads you to anticipate the next; or as you might intuitively and almost unconsciously travel along the series of landmarks that guides you to a friend's house and from there to the supermarket and from there back to home. The sky regions, for example, pass in as orderly and ordinary a fashion as the sequence of towns on a car journey, marked by welcome signs along the road. We might sense our progress northeast from Houston, say, as the rhythm of the sky realms Houston, Fort Worth, Memphis, Indianapolis, Cleveland. A flight from Arabia northwest to Europe traverses Jeddah, Cairo, Hellas, Tirana.

When I fly from London to Los Angeles, first is England, of course, much of which has already passed by before a heavy 747 reaches its first cruising altitude. Next are Scotland's great cities, Edinburgh and Glasgow, which always appear on our computer, but rarely, given their typical weather, in the window itself. Then there's Stornoway, in the Outer Hebrides—there is a bea-

con there, and to cross it feels like an aerial Land's End, the sky tip
of Britain—and then if I can see the water I may remember the
song I love by Karine Polwart, about the sea off Scotland, where
"the waves swell like a barley field that's ready to lay down." Next,
perhaps, are the Faroes, though not every route goes near them
and they are shyer, more cloud-veiled even than Scotland; I have
only seen them a few times.

Next Iceland's mountains and glaciers appear as digitized
bumps on the screen, and occasionally in white, outside the window. Then sea, and perhaps night itself this far north, depending on the season, and Greenland, which is often brilliantly clear.
After Greenland comes hours of white over Canada, a wilderness
that eventually fractures into fields and roads and other sensible or
familiar ideas. We cannot see the American border itself, but we
overfly Interstate 90, the easy-to-spot road that runs clear across
the continent between Seattle and Boston.

Then, if I am approaching Los Angeles from the northeast,
come the Rockies, and road-etched deserts, and more mountains
before the city waiting beyond, on the sea. If, however, I arrive
from the north, I can count off the landmarks of America's
snow-fired ring of prominent volcanoes: Mount Baker, Mount
Rainier, Mount Hood, the sky-blue caldera of Crater Lake, and
soon enough the snowy, stand-alone slopes of Mount Shasta,
America's Fuji, which dominates the skies of Northern California
and is said to be inhabited by the spirit of the Above-World, an
assertion that few who see it, whether from below or above, would
bet against.

Such geographic mileposts echo in the airplane, and even in
the mundanities—sleeping, eating—of a pilot's life. The timing
of our rotating breaks means that changeovers usually take place

over roughly the same parts of the earth. On many routes between London and western North America, for example, Greenland's mountains roughly correspond to the first changeover, and so the mountains, approached from the east, are to me associated with leaving the controls to go to the bunk. Similarly, they are what a pilot returning to duty expects to see crossing the windows of the cockpit when the other pilot, reminded by the sight of white peaks rising from blue sea, has gone off to rest.

Food, even, mirrors the pace of the turning world. Places in the sky take on the quality of a clearing at the bend of a hiking trail where I stop to eat lunch, marking a particular and regular point in the journey and my appetite. When I imagine Las Vegas, how it looks from the air, I think of it as a quiet spot in the sky that I associate with sandwiches and coffee, because it's where— or when—I often have a snack before the busy descent into Los Angeles begins.

I'm a student pilot, near the beginning of my flight training in a small single-engine plane. I'm flying alone, somewhere northeast of Phoenix. I'm lost.

I am navigating, or trying to navigate, visually. I have a detailed chart that depicts mountains, roads, settlements, radio masts. I match what I see on the world below to the chart, and what the chart leads me to expect next I match onto the world, back and forth, back and forth, between the two. But the afternoon has grown much hazier than forecast, and for the first time I experience one of the effects that can suddenly make this kind of visual flying difficult. I can see straight down through the haze, but anything even slightly ahead or to the sides is obscured, an effect familiar to anyone who has ascended a skyscraper on a misty day.

A pit forms in my stomach when I realize I can no longer connect anything I see through the small window of visibility below the plane with anything on the map inside the plane. I work again from the last position I was sure of, and the minutes that have passed since I was sure of it, a rough calculation that suggests a time-dependent circle I must be somewhere on. But soon this ever-expanding circle will meet mountains, the jagged, sandstone-colored peaks hiding somewhere to the northeast of me in the haze. Hardly less worrying is the strictly controlled airspace around the international airport in Phoenix, from which my intended route would have kept me well clear.

I am about to call a controller for assistance, who would give me a code for the onboard transponder that would help the controller steer me home. Then I remember that there is a navigation beacon right at my home airport, in the eastern part of the Phoenix area. My instructor taught me to use such beacons weeks ago, almost casually; it was certainly not part of the lesson plan for this kind of flying. "Just in case," he winked. I dial up the beacon. It flickers to life. I exhale, then follow the needle like a trail of electromagnetic breadcrumbs through the day's unforecast murk. I turn sharply right and descend through the haze, and soon I see the runways ahead. I rest my hand on the top of the dashboard in relief and thanks, and then I land.

How do pilots and planes know where they are? It's an excellent question, but one I am almost never asked now. The assumption is that GPS answers it, though small planes, like the one I flew in Arizona, may not be equipped with GPS receivers. Most airliners now make use of GPS. Often, it's been added onto an airliner that was not originally designed for it. There are many such technologies—related in particular to communications, and to the avoidance of other aircraft, wind shear, and mountains—that accrete in aircraft systems as layers of progressively higher neurological functions evolve in living organisms, while older systems still twinkle in the lower-down layers.

One of these older systems is *inertial navigation*. It ensures that the airliners of the world could find their way home on the darkest and cloudiest night, even if all the GPS signals, air-traffic control centers, and ground beacons fell silent.

Imagine you have been blindfolded in a stationary car. Then you feel the car accelerate to roughly highway speed. After about an hour, you might guess you were 60 miles from where you started. Then, if you felt the car turn by 90 degrees or so, before driving for another half an hour, you could draw a triangle and have a further guess as to where you were. Similar to the vestibular system in our ears, an inertial navigation or reference system senses these two qualities: acceleration and rotation.

Acceleration is measured by accelerometers, which are relatively simple devices. To accurately measure rotation, inertial systems use gyroscopes, which are anything but straightforward. Originally mechanical (a spinning top, one of the world's older toys, is a basic gyroscope), the gyroscopes in modern airliners most often use light rather than spinning discs or wheels.

The name given to such a light-based tool for measuring rotation is *ring laser gyroscope*. We think of laser beams as the very definition of straight. A ring laser gyro forces such a light beam into a closed path. Imagine a cube of glass with a tunnel drilled into it. The tunnel turns corners, forming a perhaps triangular lightcourse in the glass. At one point in the glass tunnel, light is fired in both directions. The beams travel around the loop with the help of reflectors, before meeting on the far side of the tunnel. If the device has not been rotated, then the beams arrive at the same time on the far side. But if the device has been rotated, then one beam will travel a slightly longer path through space, and arrive slightly later than the other.

A rough analogy is to imagine a round, frictionless billiards table (indeed, gyroscope means "circle watcher"). If you roll two balls away from you around the edge of this table, in opposite directions, toward a friend standing across from you, they'll reach your friend at the same time. But if you start to rotate the entire table after you have rolled the balls, one will roll for a longer distance, which takes a longer time. It will arrive at your friend later.

Isak Dinesen wrote that language "is short of words for the experience of flying, and will have to invent new words with time." The terminology of aviation is occasionally clumsy—we often speak of the brakes we use in the air as *speedbrakes*, for example, as if there could be any other kind. But the language of inertial systems is a high sort of technical verse, the engineering equivalent of Petrarch. The designers of these light-boxes speak of the *body frame*, the *local level frame* and the *earth frame*. They deal in *gravitational vectors*, the *transport rate*, the *earth rate*, and days referenced not to the sun but to *sidereal time*, the rotation of the planet against the background light of distant stars. The engineers responsible for inertial systems talk of *random walk* and *coasting*, *northing* and *easting*, and the *spherical harmonic expansions*.

There is another dimension to the poetry of inertial systems on many commercial aircraft. They require a few minutes of perfectly motionless concentration and reflection on the ground before each flight. This moment of Zen, or a sort of preflight meditation that a nervous flyer might practice, is called *alignment*. Before a system can track the motion and orientation of the plane, it must know in which direction the center of the earth lies, which it senses from gravity, and which way the plane is pointing, which it senses from the turning of our planet. If the plane is moved during this alignment period, it will display a message that says in effect: "Please, be still until I am ready."

Once an inertial system is aligned, it serves several important purposes. One is to navigate by summing up the accelerations and changes felt, as you might when blindfolded in a car. Another, less appreciated, function is to track which way is up. The attitude of a plane—the angle of its nose in the sky—is so critical to flight that it dominates the central screen, the *primary flight display*, in front of each pilot. This is the first thing I explain to guests in the flight simulator: the deceptively simple sky-blue and earth-brown horizontal division on cockpit screens shows not where we are, or in which direction we are moving, but rather, which way we are pointing (which is often markedly different from the direction in which we are moving). A plane that flies to the other side of the earth may, by the end of the flight, be close to upside down compared to where it started. Inertial reference systems keep track of what we might call *local down*, all around the world.

The intricacies these devices must grasp are subtle and numerous. When altitude increases, gravity decreases ever so slightly, and an inertial system must account for this. On a rotating fairground ride, the faster you spin, the more you are pressed against the wall; the plane follows a curved line around the earth and the inertial system must similarly account for the forces that keep it on its ever-bending path. They must account as well for the occasional gust of wind during alignment, the imperfections of the earth's sphere, the temperature of the device itself. Consider, too, that on a chessboard, instructions to move left five spaces and forward four spaces, say, are commutative—the result doesn't depend on the order in which you carry them out. But when it comes to the changing angle of the plane in space, it matters a great deal whether you rotated left, say, before or after you spun forward. An inertial system must unpick the details of the airplane's rotations carefully indeed, in order never to lose sight of which way is down.

As navigation devices, inertial systems are not as accurate as GPS. In flight they degrade further as the hours and miles pass, and small errors accumulate and snowball through their dark calculus, eventually reaching the order of miles. The 747 has three separate inertial systems. We can display on our map screen where each of the three thinks we are. Each calculated position appears as a small white asterisk, informally called a *snowflake.* I have never seen all three snowflakes in the same position. Nor are the snowflakes steady. They quiver visibly on our map of the world.

Still, even with its tremulous inaccuracies and strict meditation regime, an inertial system retains one enormous advantage. In practice today, GPS data and the aircraft's altitude are widely used to bound or limit the errors of inertial systems. But in theory, once set up, inertial systems do not need any outside source of information to know where you are, how fast you are going, and which way you are pointing. They just *know*—without looking at stars, maps, satellites, or scenery, without interrogating anyone or anything. Nor can they be interfered with externally—indeed, the development of inertial navigation was spurred by the need for accurate, jamming-proof guidance systems for missiles.

Flying over north London I can see a churchyard in which I sometimes sit with a coffee, where the tomb of John Harrison, "late of Red-Lion Square," stands. Encouraged by the astronomer Edmund Halley, Harrison developed the "sea clocks" that helped solve the longitude problem, the difficulty with determining one's east–west position at sea, an achievement so important that the officials who recognized it were known as Commissioners of Longitude. At such moments over London, as we come to the end of the planetwide countdown that every flight to this city effects, our longitude is nearly zero; it may ticktock from west and east and back to west as we cross the Greenwich Meridian in the next minutes of our approach pattern to Heathrow.

I reckon we could just about explain to an admiral or navigator from several hundred years ago how GPS works. We might say that we have essentially launched new stars into the sky, and that when we can see them, when we have a *line of sight* to them, their timed signals help us navigate. But imagine how much more impossible an inertial system would have seemed to our ancestors: a device that needs to see nothing, that you could cloak in heavy fabric, place in chains in a chest, cart across town, and roll down a hill, without its losing track of either its position or of which way is up. To our ancestors such a device—the stateliness of its sealed calculus, the wayfinding light that flickers deep within the darkened glass cube—might be more miraculous than GPS, or the airplane itself.

Before inertial systems and GPS were developed, aircraft navigators on flights over the sea, far from radio beacons, would use celestial navigation techniques to plot their position, cloud cover permitting. I have occasionally flown with senior pilots who still knew how to use a sextant. On modern 747s there is an overhead handle that we would pull in the event of cockpit smoke or fumes, to exhaust them directly to the atmosphere. (I once heard a perhaps apocryphal story of a long-retired pilot who would attach a hose to this vent in order to vacuum the cockpit.) This vent occupies what on previous 747s was a port designed for a sextant, a means of taking star sights; a hole in the airplane designed for clear nights and a bygone age in which celestial navigation was an unremarkable part of aerial wayfinding.

I have never crossed an ocean unguided by the constellations of GPS. But early in my career I occasionally flew an aircraft that had inertial navigation but no GPS, from London to Lisbon. On certain routes over the storm-mauled Bay of Biscay, the plane would sometimes veer out of range of the ground-based navigation aids on the French and Spanish mainlands. A small memo would then flash up on a cockpit screen, informing us that the plane had lost its last references to the outside world. It was now relying solely on its own internal sense of direction—it was thinking inside the box—to guide us to the far coast.

There is an arrow standing in a lake in a garden in Singapore that I walk past occasionally when I'm in the city, after lunch with a friend from childhood who now works nearby. The arrow, surrounded by water, points to England, to the observatory at Greenwich. It marks a spot chosen a century ago by surveyors of the earth's magnetism.

The maritime, navigational use of compasses in the Mediterranean dates to the thirteenth century. It is pleasing to think of how long those who have moved in the blue between cities have been guided by the simple compass, of how long this energy from within the earth has been a light to us. Some birds use the earth's magnetic field to navigate, too, and the analogy with airplane navigation appears sound—that independently, birds and humans stumbled across this unlikely gift from the earth, this unseen force that gives direction to lonely travelers and that it's so easy to imagine we might never have known about.

To the systems in modern airliners, though, magnetism is a fiction.

The distance between the magnetic and the geographic North Poles means that there are two kinds of headings a pilot may speak of: *magnetic* headings, referring to the magnetic North Pole; and *true* headings, referring to the geographic North Pole. The difference between the two, between magnetic and true, is called *declination* or more usually *variation*. Variation is not the same everywhere over the earth. In Glasgow, it is nearly negligible, at 3 degrees west; in Seattle the variation is about 17 degrees east; in Kangerlussuaq, Greenland, it is more than 30 degrees west. (Another complication is *compass dip*. The magnetic lines become vertical where they converge at the magnetic poles, as if you held a long blade of grass in your hand that was upright when it left your fist, but bent away to nearly horizontal further along its length. This means that standing at the magnetic North Pole, north is straight down, while south is directly above your head.)

Mariners were aware of variation, of course. Seafaring navigators once measured variation twice a day, at dawn and dusk, to track the local difference between true and magnetic headings. Cape Agulhas, the southernmost point of Africa that also marks the official boundary between the Atlantic and Indian Oceans, was named Cabo das Agulhas, the Cape of Needles, because five centuries ago Portuguese sailors noticed that magnetic and true north were nearly aligned here. Nowadays, pilots on a modern airliner can choose to display either type of heading. At the flick

of a small switch the whole compass *rose* on our digital map will
rotate left or right. It is a disconcerting moment when you first see
a compass, which you imagine as a deep and incorruptible arbiter
of direction, spin like a top.

Most of the time we fly on magnetic headings. The reason for
this is largely historical. In the early days of aviation, pilots—like
birds and mariners—only had magnetic directions to choose from,
because they only had magnetic compasses. And so even today,
when air-traffic controllers ask a pilot to take up a heading of
270 degrees, or due west, the controllers almost always mean not
the 270 degrees that is actually west over the surface of the earth,
but the 270 degrees that is displayed on a magnetic compass in
that part of the world.

Yet the heading display on a 747, like that on most airliners, has
no magnetic inputs. It is a surprise to new pilots, who have flown
and studied and been tested on the vagaries and inherent errors of
magnetic compasses, to realize that on a typical modern airliner
there is nothing to sense the magnetism on the earth and feed it
into the computers that generate our display of magnetic headings. There is only one magnetic compass onboard—a forlorn,
technically isolated backup device that is never used in normal
flight. On some aircraft it's even hidden away, to be pulled out
when needed, which is essentially never. It's no small irony that
the complicated electrical fields generated by the airliner's systems
themselves disturb magnetic compasses.

To display magnetic headings without using a magnetic compass, the plane consults its map of magnetic variation. The plane
knows the pilots are not using a magnetic compass, but if they
were, it knows that in this position over the earth, it would read

this. And that's what is displayed in the cockpit computers. In other words, the world's airliners fly on magnetic headings derived from a preloaded map of magnetism, rather than actual compasses. If the earth's north and south magnetic poles suddenly reversed themselves, or even if they stopped their eternal flickering one night, the pilots of commercial airliners would not see this on their screens—though birds, the pilots of small planes, and old-school hikers would notice immediately.

When I think of the long history of compasses and seafaring, or when late on autumn nights I see the northern lights, the solar wind catching in the harp-like lines of magnetism that congregate at the pole, it seems fitting that something as primeval and eerie as magnetism would have only such a ghostly sort of prominence in the shiniest machines of our age.

A further oddity of magnetism is that our elaborate fiction of it must be regularly updated. The magnetic North Pole, the star that our compasses orbit, is itself on the march—from northern Canada toward Russia, at a pace of several dozen miles per year, in a process known as *geomagnetic secular variation*. This motion means that the charts of magnetic variation must be routinely redrawn and the maps in the computers of airliners reloaded, even though nothing in the airliner's computers can detect these changes. Runways numbered according to their magnetic direction (e.g. runway 27 points roughly 270 degrees on a compass) must occasionally be retitled too, their number-names spun, and all the airport signs remade or repainted, and all the charts onboard the world's aircraft updated, to follow the latest twists in the planet's old magnetic tale.

Machine

I'm at a small airfield in rural Massachusetts, aged sixteen perhaps. It's a place that I occasionally came to with my parents when I was younger, to eat doughnuts and watch the small planes land and taxi in behind a low metal fence, the clear boundary of an airfield that many who love airplanes will have a memory of deeply wanting to cross. The planes park, the pilots and passengers get out, they walk into the lobby of the single-story building. They were in the sky; now they are here. They get into cars. They drive away, rescind a dimension, just like that.

In the lobby are a vending machine and a glass-topped counter through which I can see several shelves of maps and navigation tools for sale. On the wall behind is a bulletin board with all-capital letters stuck to it, of the type that you see in delis and diners. Here it's a menu, too, a list of the aviation services provided at this airfield, and their prices. I know these prices by heart. I've been saving from my paper route and restaurant job, and now I'm here for my first flying lesson.

It's early autumn, one of those clear, warm, bone-dry, and mosquito-free New England days, of the sort that draw people to Northern California when they realize they can enjoy them there the whole year. The leaves on the trees around the airfield perimeter are beginning to change; on the mountains nearby, as my mother would say, the color is "farther along." I greet the instructor and purchase my first logbook, navy blue, from one of the shelves in the glass counter. We go outside and walk up to the white plane. I'm surprised to suddenly find myself on the other side of the fence.

Until today I've only seen planes from a distance or entered them through a jetway that masks nearly all of the experience. I've never touched the outside of an aircraft before. There is a surprising lightness to the plane. The doors feel flimsier than those on any car. There's awkwardness, too, a sense that the plane is crafted for something other than motion on the ground or human comfort. There are plenty of opportunities to hit your head on something that looks expensive. The plane's wheels are chocked, and the wings are tied to hooks in the tarmac. There must be experts whose job it is to design airfields and this must be one of the things they think of to install—hooks to tie our wings down to, when we are not in flight.

Naturally, the instructor is wearing aviator sunglasses. He is inspecting the aircraft with the seemingly contradictory mix of utter familiarity and deferential caution that I will later associate with pilots but even more so with the engineers who check and repair airliners. I follow him as he patiently explains what he is looking for at each point in his careful circumnavigation of the plane. He takes liquid from the bottom of the wing's fuel tank, as if he is drawing blood, with a specialized tool that aspiring pilots may acquire along with their first logbook. He holds the clear column of what he has drawn up against the light, against the blue of the sky. He is checking for water. He pauses, stares straight

at me. Water is bad news, he tells me. A new fact about water.
Twenty years later I will read in my father's notes that when he
lived in Stanleyville, the city in the Belgian Congo now known
as Kisangani, another missionary took him on a flight over the
Tshopo River. Their small plane nearly crashed into the river's
reservoir because the fuel cap had been left off the night before,
allowing rain into the tank.

The instructor and I have come full circle around the craft. The
inspection is complete. He opens the door of the airplane and
smiles, gestures, tells me again to watch my head. While I'm carefully climbing into the machine, he unties the wings.

Since I have become an airline pilot, I am occasionally asked:
"What does it feel like to fly?" Perhaps the most honest answer
is that I don't know. What passengers see of the world is permanently framed by the iconic ovals, the windows cut from the hull
of the vessel. Even pilots, with their wide and multidirectioned
blessing of views, are surrounded by surfaces of intricate electronics, busy computer screens, buzzing radios—the plane and its permanent, metallic mediation of the experience of flight. Planes are
noisy, particularly small ones. I have the sensation of truly flying,
peacefully and silently as we do in our dreams, more when I'm
swimming than in any airplane.

Aside from the lines that tie the wings of small planes to
the ground, it's seat belts that are the simplest reminder of the
machine, of what, exactly, is flying. Whether as pilots or passengers, our experience of an airliner begins when we walk inside a
contrivance the size of a building. To do what we call flying, we sit
down in this. We tie ourselves to it.

Many pilots, of course, love the airplane precisely because it
is a machine. An air*craft*, a root that equals strength, skill. Nor

is it clear that passengers, either, wish to pretend the ship away. We might consider why photographs taken from the window seat are almost always more evocative if they include something of the airplane's structure—an engine, the curve of the window frame, or the lines of the wing. The airplane's photogenic presence is something more than foreground technique. Perhaps the plane stands in the imaginative place of flight itself, of the experience we cannot have directly, as if looking from the window we say: "Yes, of course, we will never fly quite like we do in our dreams. Dreams are easy; this is real."

Machines, indeed, hardly get more difficult. Pilots occasionally experience the unusual sight of an airplane indoors. A large plane inside a hangar looks much bigger still, in the same way that even a small car suddenly appears enormous and awkward in a garage. In a hangar you see the many steps, platforms, and hydraulic lifts that are needed to return an airliner's structure to a human scale, much as the hulls of boats are brought close to human eyes and hands in dry dock. In some hangars the plane is all but disassembled for checks and maintenance, as if an actual airplane had enacted its own *exploded view drawings*, the engineering diagrams in which parts are pulled away from each other, for the purpose of understanding or duplication.

It's been fifteen years or so since that first lesson in a small plane over the autumn-tinged hills of western Massachusetts. I've been flying the Airbus for some time now. I have just finished a flight to Hamburg, during which we spoke to controllers who answer to the name of Bremen Radar. From the cockpit the captain and I saw the sun-sprayed Elbe and the beacon of the same name, which was lit on our screens and receivers. My job has brought me to Hamburg many times already. So today, instead of my usual walk along the Binnenalster Lake toward a café in the city's old town, the captain and I are making a visit to the plant where the airliner we flew today was built.

We organized the visit only an hour before we left London. We called the factory and said: We fly the Airbus and later today we are flying one to Hamburg that was built there. One moment, said whoever answered the phone, then: Very well, what time do you land? We are treated well at the factory—vigorous hand-shakes, a huge lunch, a gleaming German luxury sedan—so well that we fear there's been a certain misunderstanding, that our hosts are expecting us to sign purchase agreements, to lay down an airliner-sized deposit, before the end of our afternoon together.

The plant, a complex of enormous buildings, is an exalted place—though like airplanes themselves, a mix of the extraordinary and the humdrum. The interior volume is enormous, inhuman in its scale, yet as clean as a hospital. Some airplane factories are so large that clouds once formed inside them, a foreshadowing of the sky to come for each newborn jet. As a workplace an aircraft factory seems as inherently gratifying, as backgrounded by a general radiation of satisfaction, as cockpits are. I am struck for the first time that those who work here may have a sense of the airplane that is deeper, even, than that of those of us who fly them. The workers who crafted early airplanes were once compared to those who built Chartres. Still today, in a cathedral of industry, new vessels are taking shape; great forms are raised by the most skilled hands of our age.

While each part waits for its appointment with the new machine, it may rest on a shelf in the factory. On an aircraft that is only days away from delivery, I see an otherwise complete cockpit that has no seats—a cartoon awaiting a caption, or a vision of some pilotless future of aviation. Later, on the factory floor I see a rack against the wall, stacked with little trash receptacles, each equipped with a simple flap. These bins are certified, weighed, and accounted for like almost no others. At a precise moment in the birth of each airliner, a worker will carry two of these into the unusual, newly sculpted room that is the cockpit, where each will spend the next two decades or so receiving banana peels, empty bags of nuts, pens that have run out of ink, and receipts from foreign restaurants.

A few minutes after my encounter with what are almost certainly some of the world's most expensive trash cans, we enter a large hall that is lined with enormous slices of fuselage—the double-decker cross sections of a new aircraft model. The silent sections are unpainted and separate, and no workers are moving around them, a stateliness of process, a tranquillity that only heightens both their complexity and the scale of their itinerant life to come. Though the building shines with modern machinery and lights and the sight could be taken from science fiction, I feel that I might as well be in some bygone forge or foundry, that the roar of blast furnaces must be near, that something raw and precious has been poured into new alloys, hammered, cooled into the curved walls that will now be joined.

A few minutes later we turn a corner and I see a pair of wings, as they are quietly maneuvered into position. I think of the word *airborne,* what this plane's passengers will be, that these wings will hold them. The sight, though, as if lifted from some gauzily bright future of our species, is anchored by the weight of ancient or archetypal rites—the laying of a keel, the benediction of what will bear us over the world—that we hear in "Marina," by T. S. Eliot: "The awakened, lips parted, the hope, the new ships."

As a postgraduate in England I studied African history, and halfway through my studies I traveled to Nairobi, where I planned to stay for a year. I flew from London to Muscat, and then to Nairobi. On the way we flew down the Somali coastline—I had never before seen a land of such colors, mixed yellow and deep red—and I realized that one reason for my particular excitement about this journey was that it required two flights, not one. Perhaps, even, I was more excited about flying to Kenya than about what I hoped to find in a dusty archive there.

My mother and I both loved Isak Dinesen's *Out of Africa*. My mom had left her small hometown in Pennsylvania for work and college, and later lived in Paris; her love for the book was perhaps because it's a tale of a great life-journey that starts and ends in a small place. Personally I loved the book best for its descriptions of flight, and as the plane descended over the hills around Nairobi I half-seriously asked myself whether, if the book had not contained such elegiac descriptions of aviation, I might have picked a different branch of history, if I might now be flying to a different country or continent. My mom's copy, which came from the Book-of-the-Month Club, is still on my shelf at home, though it's missing its cover. Next to it is a first edition of the book—published in 1937, the year my mother was born—that I was given not long after she died.

Each morning I walked into Nairobi from a small apartment just north of town. In the archive I sat with a notebook and a computer and worked through stacks of colonial-era documents. For lunch I walked around the center, trying different cafés. Nairobi

was first a railway camp; often I took a sandwich to eat on a bench
at the city's fabled station, where the trains from Mombasa still
come.

I left the archive early one day to visit the house of Dinesen
and the grave of Denys Finch Hatton, both in the Ngong Hills
they flew over together. When Dinesen came from Denmark to
Kenya she sailed to Mombasa on the coast, and then traveled on
the train that climbed up to the new city in the highlands. Later I
would learn about the navigation beacons named for the Ngong
Hills. The name and the frequencies, 315 kHz and 115.9 MHz,
are marked near a peak on the modern arrival charts that every
airliner flying here will carry. My tour's van left from an international hotel in the center of the city. In the lobby I saw happy
crews from the airline I would one day work for, arriving from
their flight, suitcases clattering over the floor, lining up to organize
other tours. Their work, I marveled, had been to fly here.

A few months later I decided to leave—the library, Nairobi,
my postgraduate program. At that point I was still not certain I
wanted to become an airline pilot. But on the flight back to London I asked to visit the cockpit and, during my half-hour visit,
we overflew Istanbul. The Golden Horn and the Bosporus and
the domes and minarets of the city were lit sideways, perfectly,
by late-afternoon sunlight. Asia was disappearing under the nose;
there ahead lay Europe; here, between them, it must be Istanbul. "The city of the world's desire," said the captain, pointing.
"Constantinople," he added, when he saw my blank face. I was
struck by the age of the Bahraini copilot, who must have been
only in his late twenties. He was hardly older than me. We talked
about his career, my interest in flying. "Oh, you should do it,"

he said with a smile from behind his aviator glasses, in barely accented English.

I returned to my seat, put my headphones on, and watched Europe unroll from one end to the other. The music's effect on my reflections and the passing world was even more montage-like than usual. I no longer keep a regular diary but when I was younger I loved to write in the window seat with my headphones on. I still occasionally see passengers who have made this same arrangement of window, music, and paper—travelers who have taken an old-school approach to their geo-graphy, their writing of the world. That pilot was right, I thought. This is what I love best. The moment I made my decision to become an airline pilot came a few hours later, when I was on a bus, rolling clockwise down the M25 motorway from Heathrow.

My journey to the cockpit had one more stop: a job in the business world to repay student loans and start to save the money I expected I would need for my flight training. I had often heard complaints about the amount of travel that management consulting required. Naturally I applied to every consulting firm I could find. I had no response from the five large consultancies I first applied to. Later I found a glaring spelling mistake in the first line of my cover letter. I corrected it and applied more widely. Eventually I took a job at a much smaller firm in Boston; I was drawn to its friendly atmosphere, the possibility of travel all over the world, and an office in a charming redbrick dockside building that enjoyed stunning views of the city's old harbor and the airport across it. Three years later, I left that job to start my flight training in Britain, on a course sponsored by an airline, part of a group of aspiring pilots who are today still my best friends at work.

As an educational experience, private flying is practical and hands-on. Commercial flying, on a residential course like the one I followed, is entirely different. About half of the eighteen months or so of that course were spent entirely in the classroom. And so, having left the academic world behind some years earlier, I was surprised to find myself again at a desk with a notebook, worrying about exams, and studying late into the night with friends in the common areas of a crowded hall of residence.

The historian I. B. Holley wrote that we have neglected the creativity that makes our technology possible. I certainly had. When I returned to the classroom I held the simplistic view that academia, and perhaps much of the thinking and working of the world, was neatly divided. There were so-called creative or "soft" fields—those whose practitioners try to work outside the box, or to think about the biases of a box, who talk about why boxes are important or beautiful, and the history of boxes, and why some boxes hate other boxes, and how boxes are depicted in the arts. Then there were the "hard" professions—those whose acolytes are devoted to the evolution of boxes, to their chemistry or mathematics, or to the design and construction of more reliable ones.

Never in my life has a view that I held been overturned as cleanly and quickly as this one. Within the first hours of my new classes on aircraft technology, and again and again throughout my training, I was struck by the extraordinary creativity of engineering, by the art of flying: how connections are made between substances or disciplines, how an effect in a system is conjured as carefully as that of a story, a poem, or a song. And engineers are confined by a frame of physical laws and, especially in aviation, by a web of rigid constraints—weight, reliability, near-perfect safety—that might make a composer of haikus blanch.

I marveled, too, at the similarities between engineering and biology, how engineers are the agents of a kind of evolution, the conscious evolution that is the work of an industrializing species. This thought first occurred to me early in my flight training, when I was taught about a device called a *fuel-cooled oil cooler.* The oil in the engines gets very hot, while the fuel in the wings gets very cold, especially on long flights at high altitude. So airliners may deploy heat exchangers, which allow the engine oil to transfer excess heat to the fuel, without mixing the two. While the instructor was describing this, I thought of the high-school biology class in which I learned that whales use a certain kind of heat exchanger to transfer warmth from arterial to venous blood. It's just one of many examples of the convergence of evolution and engineering. Airliners have skins that like any skin serve to regulate what passes through; they have circulatory systems; they maintain a self-regulating, all but biological level of homeostasis. Like our awareness of the location of our limbs, they have proprioception of the positions in space of flight controls. Airliners' ability to self-monitor many systems, and their carefully graded hierarchy of notifications and alarms, have many features in common with pain. Airplanes store maps and adapt them in real time; they sense many qualities of the world around them such as temperatures and wind, and the presence of land below or precipitation ahead.

When pilots arrive at the threshold of an airliner, the airplane is almost always already powered, lit, and air-conditioned, drawing electricity either from the airport—the jet plugged in as simply as the toaster in your kitchen—or from an auxiliary engine in the tail.

Sometimes, though, an aircraft that spends the night at an airport is completely depowered. On certain early icy mornings—in my memory such mornings have always been in northern Europe, in winter, when we reach the plane well before the first hint of sunrise—I have walked up to the jet and opened the door slowly, moving its surprising weight aside to the locked-open position. Inside there is total silence, the stygian feel of the inside of a snow-covered car, and the same sense that the vehicle is cold and unprepared for the scale of its intended motion.

I then make my way to the dark cockpit, to begin to work through the airplane's first checklist by flashlight. This activates some of the aircraft's most critical functions, those that are hardwired to the batteries. They are the first systems the plane powers and the last things it would relinquish, as a body prioritizes blood to the brain. It is as though we are discovering an alien spaceship, perfectly functional, and slowly, line by line from a manual, re-conjuring its evident brilliance centuries after it was abandoned.

Next I start the auxiliary engine at the back of the plane. It takes only a minute or two, but it always feels much longer; I do not have many attempts before the battery would be drained entirely. Success, when it comes, is a series of auspicious flickerings in the cockpit and down the passenger cabin, as systems activate, lights turn on, cooling fans begin to whirr, screens flash and go blank, bleached colors appear on them and slowly turn true. Many components begin to test themselves; warnings arise, then quickly clear. Electrons begin to flow through the nerve wires, hurrying light to the distant wingtips or returning with news of the quantity of fuel onboard or the present outside temperature, as the plane awakens to its purpose.

My father's geographic circumstances—born and brought up in polyglot Belgium, sent to work as a missionary first in the Congo and then in Brazil, finally immigrating to America—meant that he was required to study and work in a variety of languages. The way he spoke of each—their intricacies, joys, and eye-rolling quirks—was similar to how pilots talk about various aircraft types they have flown.

It's often assumed that an airline pilot can fly any kind of airliner. Pilots typically take a set of exams, both in classrooms and in the air in small planes, to obtain a series of licenses that culminate in a general air-transport license. Then we obtain a *type rating*, a separate license to fly one specific kind of aircraft, such as a Boeing 747, or perhaps a series of aircraft designed with similar cockpits for exactly this purpose. A type rating involves a course of several months, including classroom and simulator training, as well as actual flying. When we switch to a new aircraft, the new type rating replaces the old one, and usually we are no longer permitted to fly the previous type. Some pilots fly a dozen types or more in their career. I may fly only three—the smaller short-haul Airbus airliner I started on, the Boeing 747-400, and probably one new type, between the 747's retirement and my own.

Much of the day-to-day professional knowledge we must maintain is specific to our aircraft type. We spend much of each day, or night, inside it; when we sit down, it will feel like a second home. Our connection to this aircraft will even color our experience of travel as a passenger. When I fly as a passenger on the Airbus, which I flew before the 747, it has the familiarity that alienates, like walking past a restaurant where you broke up with someone long ago. In contrast, when I fly on a 747 as a passenger I feel a peculiar comfort or satisfaction that is something more than knowing what the various noises mean.

The bond between a pilot and their current type of airplane

is hard to pin down. Language, as my father's sense of languages reminds me, is perhaps the best analogy. Indeed each aircraft type or family has its language, or at least its own dialect, and analogous devices and procedures often have different names on different aircraft. Acquiring these words and their correct usage is a significant part of the work we put into a new type rating. In a phenomenon called *type reversion*, a pilot inadvertently refers to a term or procedure from a previous aircraft type. There is a friendly rivalry between the pilots of Boeing and Airbus aircraft, which in addition to everything else are two competing realms of language. On the Airbus, the fully stowed position of the flaps is called *flaps zero*. On the 747, the same position is called *flaps up*. Once, soon after I switched from Airbus to Boeing, flying with a senior captain, I mistakenly asked him to select flaps zero. Before moving the flaps he turned to me, with a clearing of the throat and a smile—from over the glasses resting halfway down his nose— that said: What are these youngsters coming to?

In terms of technical knowledge, a type rating is not nearly as permanent or deep a distinction as a specialization in medicine, but perhaps it's similar to the study and practice of a particular technique of surgery or imaging within a specialty. Law may be analogous, too, in a country divided into different jurisdictions that may require separate licensing—individual states, for example. Emotionally, a pilot's relationship to their type is perhaps similar to how some people respond to a prized car they have owned for a decade or two. But different cars are not as different to drive as different airliners are to fly, nor do they exclude other cars from your driving life.

Many pilots don't get to choose their particular aircraft. They may even work for an airline that has only one type, for example.

But in many airlines, pilots have some choice as to what airplane they fly, an opportunity that often arises when their company orders a new aircraft or retires an older model.

When a pilot needs to express a preference, perhaps the weightiest consideration is the distance a plane typically flies. Some pilots prefer shorter flights because they find the busy starts and ends of flights the most professionally satisfying, and the shorter each flight, the more takeoffs and landings pilots will perform. Such pilots may also fly more short round-trips that bring them home each night, rather than to hotel rooms far away. "I'm a proud flat earther," I've heard more than one pilot joke, to emphasize that they would never give up short-haul for long-haul flying.

Pilots tend to like powerful planes. I've often heard complaints about one long-retired aircraft type that pilots felt was underpowered; the joke was that it only ever got airborne because the earth eventually curved away beneath it. In contrast every pilot I've talked to who has flown the Boeing 757 has mentioned, unprompted, how powerful its engines are. But equally often I hear wide-eyed pilots marvel at the efficiency of a new airplane, after they compare the amount of fuel burned between an older and a newer, more efficient aircraft on the same route.

The differences in the cruising speeds of airliners are small. Still, some airplanes and their pilots spend their hours in the sky habitually overtaking others. It feels good—how could it not?—when you are pulling ahead of other aircraft even while maintaining your most efficient speed.

Size is a more complicated question than speed. On a small plane there may be two pilots and three or four flight attendants; on a long-haul airliner like a 747 there may be four pilots and fourteen or more flight attendants. It's much easier to get to know colleagues when there are fewer of you; and on a larger plane not only are there many more crew but there is also the matter of the greater physical distance between the pilots and those who are busy working far from the cockpit. Smaller planes can also feel more maneuverable, sportier. I once asked a pilot who flew a small regional jet how he liked his aircraft. His eyes lit up; it was, he said, better than surfing.

Still, it's my impression that more pilots prefer longer routes, and therefore the larger planes that typically fly them. One reason is the chance to see further-flung cities and countries and to flee your home weather, or indeed the entire season of your home hemisphere, for something more to your liking. Long-haul pilots also tend to have more free time at their destination, because the amount of rest required is greater when a flight is longer or crosses more time zones. And a nearby city linked to yours by small planes may be fascinating to you, or it may not be very different at all from the world you already know well. But a city that calls planes to it from far across the planet must be in some way globally prominent—particularly beautiful or beloved or enormous.

I enjoyed my years on a smaller plane. But among those short flights I always liked the longest ones best, and I knew that I wanted to spend at least a portion of my career on a large aircraft. I've been stuck since I was a child, I think, on the idea of flying far, over varied landscapes, to the biggest cities on earth. Saint-Exupéry is often credited with saying that he flew *"car cela libère mon esprit de la tyrannie des choses insignifiantes,"* because "it releases my mind from the tyranny of petty things." Releasing my mind from traffic or the line at the bank is easier, certainly, when I know that in the coming hours a quarter of the world, many distant countries of cloud, will move across the windowpanes.

Some pilots have the opportunity to try both short-haul and long-haul flying during their career, to find what they love best, or to follow their preferences as they change throughout a long career. The luckiest pilots will fly an aircraft or aircraft series that covers both short- and long-haul routes; within a single month they may experience a wider variety of routes and places than many pilots will in a lifetime. Flight attendants, too, have training specific to aircraft types, but they often hold several such certificates at a time. This means they can fly to the destinations covered by multiple types of aircraft; in this way their world is much larger than that of any of the pilots they fly with.

Some pilots joke that the appearance of their plane does not matter to them, because they are looking out from the inside of it. Still, the aesthetic qualities of airplanes are a regular topic of contemplation and conversation. Pilots might say that one airliner looks right, or that another looks—vaguely, but definitely—wrong. Or that one plane looks as though the engineers kept sticking bits on, seeking a frustratingly elusive aerodynamic solution, each design amendment then requiring another; whereas other planes look good from the start. Pilots will often remark on a new plane when they see it for the first time, puzzling over whether it looks awkward only because it's new, or because its appearance is genuinely unfortunate. We may ask an older colleague how an old and much beloved plane looked to them when it first landed decades ago.

Often a manufacturer will lengthen or shrink an existing airplane type. Aesthetically, a lengthening is generally an improvement, while a foreshortening is risky. Imagine the leverage that's added by the long handle of a tool—a screwdriver, for example, that you use to pry off the lid of a can of paint. Similarly, the longer a plane, the longer the *arm* the controls on the tail of the plane can act along, and therefore the smaller the required size of the tail. This is one reason, if a plane is shortened, the tail may not shrink along with the fuselage; it may even grow, and look markedly ill at ease.

Occasionally one airplane catches the imagination of pilots and cabin crew, or even of the general public. More than a few colleagues told me they decided to learn to fly only because they wished to fly the 747. I am never surprised when a colleague's e-mail address contains some version of those famous numbers. I occasionally go to an exercise class near the hotel I stay at in Vancouver—exercise is sometimes the best antidote to long-haul travel, whether because it resets the body's clock or only tires you out into sleeping better, I do not know—and the instructor will often sing out, at the start of a pose in which we are lying on our stomachs but lifting all our limbs: "Lift your arms, lift your shoulders, like a 747 taking off."

Recently I was taxiing a 747 past a portion of the tarmac at San Francisco that was closed off for reconstruction. More than a dozen airport workers, though presumably already accustomed to the sight of airplanes at close range, nevertheless put down their tools to photograph us. On one summer evening when I was flying near sunset over the Netherlands a different aircraft type passed over us, and the other pilot let out an aerial catcall to our 747, a low whistle over the radio, then: "I hope you have a lovely day on that lovely aircraft."

Partisans often say that the 747 jet "just looks right." I agree, but this isn't necessarily what you'd think of a plane with such an unnatural bump (a design that moved the cockpit upward and back, to permit an up-swinging cargo door to be fitted to the nose). The lines of the 747 may be so satisfying not despite this nose

bump but because of it. Perhaps it recalls a natural relationship—
that of the head of a bird, a swan perhaps, to a long body and
wide wings. Joseph Sutter, the 747's lead designer, was drawn to
birds as a child—eagles, hawks, ospreys. He might be pleased to
know that his achievement has come full circle, that a writer on
the wildlife of Virginia has described the great blue heron as the
"747 of the swamp."

Other differences between aircraft are so small in the context of such earth-crossing, mile-vanquishing vessels that it feels
ungrateful to dwell on them. Airbus cockpits are beloved for their
foldout tables, an enormous enhancement to the pilot's quality of
life when completing paperwork or a meal; I also found the cup
holders and sun visors were more intuitively located on the Airbus. Some planes have windows that open, a blessed feature when
you're dining in the cockpit between flights and wish to feel the
breeze on your face, especially if you have flown from somewhere
cold to somewhere warm and have only three-quarters of an hour
until you must fly home to winter. Some airplanes have a bathroom inside the cockpit; for this reason the 747 is often called the
en-suite fleet. (When I first started to fly 747s, the cockpit lavatory,
a standard airplane fitting, contained a most unlikely feature: a
baby changing table that was only later removed to save weight.)
Many long-haul planes have pilot bunks. On some airplanes you
have to pass through the passenger cabin to reach the bunks or
lavatories; on others, like the 747, you need never leave the cockpit
area and can move freely between the bunk and the bathroom in
your pajamas.

The best proof that the temperature outside is really as polar as
the cockpit gauges indicate is the floor of the cockpit. It can be like
ice. Some aircraft have foot heaters and some do not. When I flew

Airbus jets that were not equipped with them—my understanding is that they are an optional extra, like those a car salesman might offer to throw in during the last minutes of negotiations—I would sometimes wear heavy socks for unusually long flights. I would be in a hotel in Bucharest, in the baking height of a continental summer, thinking of the sphere of cold above even the warmest times and places as I pulled ski socks onto my feet. The 747 has foot heaters. The frozen surface of the Arctic Ocean looks better—everything looks better—when your feet are warm.

Aside from foot heaters, new technology plays a perhaps unexpected role in the preferences of pilots. When I worked in management consulting, I had the sense that everyone wanted the most advanced tools—laptops, projectors, phones. Planes, like computers and smartphones, differ in the level of technology they incorporate. Some pilots are early adopters, gravitating to the newest equipment. But it's quite common for pilots to strongly prefer older aircraft. One reason is that in such aircraft, in which fewer tasks are automated or computerized, many pilots feel closer to the simplest mechanics of flying and an older ideal of their profession. Each new generation of aircraft lays down another stratum of technological sediment between the modern pilot and the Wright brothers, and the pace of technology is such that some pilots may fear that once they leave a more traditional aircraft type, they will never again have a chance to exercise their skills in the same way.

When visitors clutching the latest smartphones come into the cockpit of the 747, they are often so shocked by its relative antiquity that they can't help but comment on it. Many pilots take such a reaction as a compliment, and joke that "it's a classic" or "it's steam-driven but we like it that way," while resting their fingers affectionately on the four stilled throttles.

*

If the now-familiar form of an airplane still holds the modern eye, it's perhaps because it holds opposites.

The routineness of air travel today, the sometimes-weary casualness with which many passengers fly, contradicts the physical grace of airliners. Yet in science-fiction movies, when the music rises and we glimpse a craft that is more poetry than machine, a shimmering vessel perhaps without an obvious means of propulsion, it is the cultural and visual lines of airplanes that filmmakers call upon, rather than actual spacecraft, most of which have no need to be aerodynamic and are therefore unattractive.

There is also the size of an airliner, set against its breathtaking reserve of speed. A large airliner, the consummate elider of place, itself possesses the scale of a structure or enclosure we might work in or inhabit. Sutter, the 747 designer, remarked that his airplane was "a *place*, not a conveyance," one that an architectural magazine would describe as the most interesting edifice of the 1960s and that the architect Norman Foster would name the twentieth-century building he admired most. Yet this building, this place, moves nearly as fast as sound itself.

Then there is the airplane's solidity, the metal heft of it, so incompatible with the ungraspable medium it moves through. We speak of a jet's weight in shorthand—340 today for takeoff to San Francisco, 385 to Singapore tonight—and I am occasionally shocked to recall that the unit we do not bother to append to these numbers is metric tons. The 747, whatever its abilities to make light of the planet, is too heavy to stand on the tarmac of many of the world's airports.

A parked airplane also embodies contrarieties of place. At the airport gate a plane is immobilized, Gulliver-like amid the vehicles, personnel, and activities that surround it. Yet it retains something of the imaginative shadow it cast when it vaulted seamlessly from Singapore to London—on the Andaman Sea, Delhi, Kashmir, the snowy peaks of Afghanistan, the holy city of Qom, the Black Sea, Transylvania, Vienna, the banks of the Rhine, the cathedral of Antwerp, the lanes of the Channel. The stillness of a parked airplane holds all places; such groundedness suggests only its opposite. That, I think, is what I realized when as a kid I saw that plane from Saudi Arabia parking at Kennedy Airport, and what I came across again, in a limited but more close-up way, when I walked around a small plane before my earliest flying lesson.

That first personal experience of inspecting an airplane I would help fly, as a wide-eyed teenager following an instructor around that small plane, has proved to be a constant in my flying career. Even for the largest airliners the tradition remains that before each flight one of the pilots will descend to the tarmac to inspect the aircraft exterior. This is colloquially called the *walk-around*.

The walk-around reminds me that the built world is composed of a hierarchy of machine-based, increasingly superhuman-scaled spaces. We know this, but we so rarely have a chance to consider this hierarchy directly from our appointed places within it. The walk-around is an opportunity to cross these boundaries. The pilot leaves the more or less friendly public part of the airport, with its windows and music and chairs and cafés, and transits both vertically and conceptually down to the working area of the field. The actual descent often takes place on metal stairs attached to the side of a jetway. The angle of the staircase may change as the jetway rises or falls, and so they are often at their most precipitous when reaching up to the high doors of a 747. These narrow and vertiginous steps—even when it's not dark, wet, and windy out—are the only ones in the whole world on which I'm religiously careful to use both handrails.

Rising noise marks the descent as steeply as the steps do. Sound floods over us the moment we open the heavy door to the stairs. Outside it's a thunderous world, even when we're wearing the required earplugs. People have no purpose or pleasure here, except to perform a specific task on an expensive and noisy machine, usually with the help of another expensive and noisy machine. From the stairs we move over the ground as we would cross the chaotic streets on our first day in a foreign city, relying on extreme caution, not the rules of traffic or the goodwill of drivers, to ensure our safety.

The area around each gate is carefully marked, in paint as well as in the mental maps of all who work here. Inside this area, certain vehicles and people move relatively freely. The border of this area, then, marks another transition, to the taxiways. The taxiways are nearly a no-man's-land, a world scaled not only for enormous machines but for those on the move. A pilot will often walk near the border between the gate area and the taxiway. If you have ever stood next to a wide, racing highway, you will know the same vague malaise of unbelonging—the feeling that you are only narrowly separated from a realm of bigger, faster, harder creatures, the opposite feeling of walking down a small European street. The teams that push back the plane are among the few who walk on the taxiways, and there are elaborate rules to protect them from moving airplanes.

The taxiway—windswept, hard, vast—is alien in another sense. Here, it's not tumbleweed that suddenly rolls past but a 230-foot-long, 400-ton aircraft, engines roaring. The passengers on planes taxiing out for takeoff have already left the humanly scaled world of the airport; indeed they have left the city, they have departed in all but the most physical sense. Faces you can hardly see in the blur of ovals provide the same flickering sense of others' lives as you get through the window of a subway train that briefly parallels your own; the sight of someone already gone, the presence of absence.

Back inside the gate area are many machines that do not fly. It becomes clear on a walk-around why toy airplane sets so often include many of the ground vehicles, the enormously varied ecosystem that swirls over each aircraft like a reef. These vehicles and the staff they carry are busy here and now because the airliner will soon be inaccessible; they are doing what cannot be done later in the sky, which is to say everything. The term for an aircraft with a technical problem that prevents it from flying is *AOG*, for *Aircraft On Ground;* a term that precisely reflects the importance of minimizing the time between landing and takeoff.

There are the trains of baggage containers, and the vehicles that load these into the plane; there is the tug or tractor—the necessarily heavy vehicle that pushes the plane away from the terminal. It is typically locked onto the nose wheel several minutes before departure, a steaming cup of coffee waiting for the driver. Most airliners, unlike almost every other kind of vehicle, cannot move backward on their own. This small but necessary reversal, the need to push a plane backward a few hundred feet before releasing it to move forward 6,000 miles, still strikes me as curious, as if the motions of airliners over the planet were as simple as that of toy planes that must first be pulled back along the floor.

There are the catering vans, lifting high on their scissor-legged platforms, ready to deliver the meals you will eat hours and miles from here over some far country of cloud; there is the refueler, pumping 25,000 gallons or more of jet fuel into the wings, most of which will have been consumed by the engines before your pre-landing breakfast is served. Engineers may have parked their airport-confined cars nearby while they conduct their checks or repairs; other vehicles carry teams of cleaners, bags of blankets recently arrived or about to depart. One vehicle delivers water to the plane; another, sometimes referred to as the *honey wagon*, removes waste; one more may be rising skyward to scrub the cockpit windows or to wash ice from the wings.

The exact route of the walk-around is rigorously mapped in manuals. I begin near the nose, which is so high that I must move far ahead of the plane to see it. To view the plane head-on is to experience the aircraft as the air itself could be said to. From the front, an airliner looks like an animal—the cockpit windows like eyes, the cone like a nose or beak. A plane looks like a bird if you account for the wings, like an orca if you do not. The zoological imagery is reflected in both versions of the terminology used to direct planes as they push back from the gate—American controllers sometimes say: "Push, tail south," while in much of the rest of the world the same instruction would be: "Push, face north." Around the nose are probes that poke out and bend forward into the slipstream. They sense pressures, help calculate airspeed and altitude; their jaunty angles, their determined embrace of the slipstream, suggest nothing so much as a dog with its head out of a car window.

The cockpit windows embody both technical rectitude and the more human aspects of aviation. Drone aircraft, as the poet James Arthur reminds us, typically have no need for windows, a disconcerting facelessness that perhaps more than their perceived autonomy explains why drones so often look like something out of a horror movie. On the ground at night, with the cockpit screens and the lights turned down very low so the pilots can see out clearly, the cockpit windows of a taxiing plane form blank panels, as dark as pupils. Before pushback, though, brighter lights may be on in the cockpit. Sometimes when I see the nose of a parked aircraft from a terminal, I marvel at its smoothed technical precision, and then I'm struck by the sudden sight of faces within it, the pilots in the windows, one turning or smiling to the other behind the thick glass. So I try to imagine this view of the pilots in flight, high over some distant land—the inaudible conversation, the cups of tea rising to lips behind aquarium-thick panes.

The very first item on the checklist to be followed in the event of damage to a cockpit window—to make sure that our seat belts are fastened—is one that seems hardly necessary to have put in writing. The windows are heated to prevent ice from forming and to soften them, in order to better absorb the impact of birds. Such multilayered panes are a reminder of the days of open cockpits, and the sophistication of the facades we now so routinely craft to halt everything—birds, snow, hundreds of miles an hour of wind—everything but light.

Though the convergence of the lines of the plane naturally draws the gaze of an observer toward the nose, it's the wings that dominate the experience of the walk-around. The word *wings* still retains echoes of the divine, as if their simplicity and beauty might lead us to forget that we ourselves make them. We sculpt them and then we fuse them to a bus. There is only one pair of wings, of course, thanks to the French aviator Louis Blériot, credited with the creation of the first practical monoplane. Embedded in the wings, in the curve of the 747's take on Blériot's innovation, are powerful lamps known as landing lights. We might thank Blériot, who first made car headlights practical, for these as well.

Whenever I look at a wingtip I like to think of the engineers and the years devoted to this pointed conjunction of design and air, where the wing gives way to the medium that breathes it to life. Such an apex should be marked with light, and so it is. Navigation lights, red and green, are arranged on wingtips as on the sides of a ship. A section in a manual describes the many exterior lights on the aircraft; it is a page I think of whenever I see blinking lights on top of radio masts or wind turbines or skyscrapers, how we mark the bodies and the endpoints of our creations.

On some aircraft there's a white light on the wingtip, visible from the passenger cabin; it catches the eye like a bright star that rises up on takeoff, to shine with us through the night.

Just before takeoff on your next flight, let your eye mark the wingtip's position—perhaps with the help of such lights—on a windowpane. What happens next is actually easier to observe when you can see a window but are not sitting adjacent to it. The wing starts to work even at low speeds. As the plane accelerates, the wing begins to rise. It works its magic first on itself—and the tips, where the wing's labor vanishes into the wind that has conjured it, lift the most. Long before you are airborne the wings are claiming weight—their weight, your weight—from the wheels and the earth beneath them. It's right to say that wings "soar." They soar and pull us up. On many planes a line drawn between the tips of the wings in flight would pass well above the fuselage, which hangs in the bow they form.

To walk under the wing is to square this upper moving marvel with its ordinary and static underside. The first surprise is the length. Passengers walk down the inside of the fuselage, but never from one wingtip to the other. The wingspan of a 747 is not far short of twice the distance covered by the first flight at Kitty Hawk. Such a structure, from underneath, is broad and wide enough to shade me, or to keep me dry if rain or snow are falling. Though sometimes, even on a hot day, there is fuel in the wing that has been deeply chilled during its previous flight—a *cold-soaked wing*. The wing may then shower melting frost on my cap or face. It has brought down the cold of somewhere high and far.

Planes moving on the ground often remind me of large seals dragging themselves over a beach, in contrast with how elegantly they glide through water; or Olympic divers, when they heave themselves from the pool and clamber up a ladder, the inevitable tedium or inelegance that bookends their moments of grace. Underneath the wing and fuselage is the landing gear—what the plane stands on, when it must stand on the earth.

Poets and engineers alike have remarked on the Wright brothers' background in bicycles. At some airports, staff use bicycles to travel around the tarmac. Often I see one of these airport bikes parked, resting on its kickstand in the shadow of a 747, which with equal nonchalance is resting 350 tons on its eighteen wheels. Later, in the cockpit, I find myself thinking about my brother and I realize it is because I saw the bicycle, and I think of the latest bike he's made for me; or I wish I had taken a picture of the one beneath the airplane, our two passions as proximate as they were for two brothers in 1903.

Consider what happens when engineers face any decision that affects the weight of an aircraft. Let's say, for example, that designers would like to install more substantial, homelike basins in the bathrooms—basins that happen to weigh a little more than the usual ones. Such a seemingly minor increase in weight in one small area may echo throughout the entire aircraft's design. The heavier basins may require slightly stronger (and heavier) structures for the surrounding walls. To carry and maneuver this extra weight may require stronger (and heavier) wings and engines that burn more fuel. Such a dramatic rippling of compromises and consequences throughout the airplane is sometimes described as a *gearing effect*. By one calculation, the addition of 1 pound to an aircraft's basic design results in a 10-pound decrease in the payload the plane can carry across the world.

I like to think that one reason airplanes are so elegant is that, as with the exacting demands of aerodynamics, such a severe gearing effect acts as a kind of natural sculptor, a scalpel on the excesses that crowd on less weight-critical human creations, the excesses that we do not know to miss. The gearing effect also suggests the great importance of anything that is permitted to be heavy or obviously ungainly on an aircraft, such as the landing gear. The enormous metal stalks of the 747's main gear legs, each as thick as a young oak, are an image of shameless brawn at the intersection between air and ground. The gear must bear much more than the weight of the plane; it must bear the impact of landing—in this sense it is an enormous shock absorber—and yet in the event of unusual stress it must break cleanly from the aircraft. It must hold the wheels and the heavy brakes, and allow them to cool. Yet even this heft swarms with sober intricacy, a wiry cloud of technical brilliances, hydraulic arteries, and the joints and appendages that allow the structures to raise themselves up at the flick of a cockpit switch so that a Swiss clockwork of complex paneling can close over them.

The tires themselves are so comically large—around 4 feet in height and a foot and a half in width—that perhaps only a child would size them correctly in a sketch. Each 747 tire may be rated to bear a load of 25 tons, as much as the monstrous tires of some earth movers, which do not have to land or hurtle down a runway on them. Often the speed limit of aviation tires—235 mph, for example—is written directly on them, along with "AIRPLANE," as if to warn against the insult of their installation on a less exalted sort of vehicle. It is hard to imagine these wheels later, unchocked, unleashed, blurring to the speed of takeoff. As the wheels are retracted the brakes will bring this rotation to a halt, so much turning turned to the heat that will be carried for many hours across the high cold sky. The end of the flight brings the sudden return of speed. The wheels are not turning when they hit the ground, but must be spun up once again at touchdown, more or less instantly, to the speed of the flying earth. Long after parking, the rubber of the tires is often still warm to the touch.

Walking around the aircraft wheels, I feel occasional gusts of the last flight's heat, its enormous, braked speed, drifting off in the breeze. The shock of a jet engine, in contrast, is that it is already so cool. We may not often think of engines in day-to-day life; perhaps we take them for granted or find them dirty or low, as if they were a brief necessity during a former stage of history, that we had no choice but to cross in order to reach this age of information. But even now, in the realm of endeavor that we have named *engineering*, aircraft engines—comparably unconstrained by cost, sculpted by air—are among the most impressive creations. Clean-lined tubes of enormous, refined power, hanging from the stately wings of airliners, the everyday word *engine*—*ingenium* in Latin, meaning talent, nature, clever contrivance—catches in the light of its origin.

Picasso—one of whose paintings would be onboard an air-

liner lost off Canada in 1998—used to address the French artist
Georges Braque as "My dear Wilbur," in an affectionate reference
to the imagination and artistry of the Wright brothers. Marcel
Duchamp, at an early exhibition on aviation, famously turned to
Constantin Brâncuşi and said: "Painting is finished. Who can do
anything better than this propeller? Can you?"

Aircraft propellers are beautiful things. It was the Wright brothers who realized they should be understood not as aerial versions
of maritime propellers, but as rotating wings (indeed, airplanes
and helicopters are sometimes distinguished by the terms *fixed-wing*
and *rotary-wing*). Yet propellers have their limitations. The tips of
the blade spin faster than the inner portions, a consequence of
physics that explains the effectiveness of salad spinners. But when
propellers get very large and fast and the blade tips approach or
exceed the speed of sound, their efficiency declines dramatically.

I have always been more fascinated by jets. To watch jet engines
in flight is still a treat to me in the passenger cabin; especially from
the rear cabin of the 747, where the scale of both the engines
and the wings is most apparent. On the largest versions of the
777 the engines alone are comparable in diameter to the fuselages
of many airliners. They conjure the speed that gives the wings
life, that gives us flight. Yet they work without apparent motion
or effort, unless you can see the turning fan, or unless you see the
light from the setting sun fall on a portion of the wing, flickering
and scintillating after it has passed through the churning column
of thrust behind an engine. A jet engine at takeoff looks more
or less the same as one shut down on the ground or sitting in the
corner of a factory; a grace of engineering and air so purely channeled that the mechanism itself is all but unseen.

The engines are identified in our manuals: *three rotor axial flow turbofans of high compression and bypass ratio.* First comes the gyral memory of watermills or crank-started motorcars, when we begin to *turn the engine.* Then, when the *start cycle* is complete, and the engines are *stabilized at idle,* we *advance the thrust levers.* At the front of the engines is the alloyed carnation of the fan. The blades stand as perfectly round as the dashed markings around the edge of a railway station clock. On airliners smaller than the 747 it is easy to reach the blades to check for ice, fingering their surprising backs, the unseen surface of the gleaming spokes. The excited five-year-old in me may give this fan a casual, affectionate whirl, and it is difficult, too, to believe how easily such a vast wheel can turn. The blades themselves are cool to the touch and, despite the name, are not sharp. On one version of the Airbus that I flew, the blades of the fan rattled when I spun it. Only at high speed would the blades hurl themselves out like the riders on a fairground ride, summoned by rotation to their appointed positions.

The underwing location of the engines on most jet aircraft gives rise to one of the more curious aspects of flying them. Imagine a cardboard outline of the plane, loosely fixed to a bulletin board with a pin, free to rotate. If the engines are below the plane's center of gravity—i.e., if they are hanging from the wings—then when power is added, the plane rotates around the pin. The nose rises, and the tail falls. This effect, the *pitch-power couple,* results in some counterintuitive flight maneuvers. For example, when we abort a landing and climb rapidly upward, we add power and pull back on the control column to bring the nose up. But as the thrust increases, as the engines *spool up,* the pitch-power couple quickly becomes so strong that we must reverse our inputs and start to push down on the control column even as we wish to continue climbing. This need to steer against power changes feels something like the steering torque that sometimes pulls a powerful car to one side when you accelerate rapidly. On some newer airliners the flight computers counteract the pitch-power couple automatically. Pilots on these aircraft must then in effect set aside one of the more eye-opening aspects of flying underwing-engined aircraft that they went to some trouble to learn.

From behind the engine you can see the *core*, the engine within the engine. *Core* is the right word for these hidden and essential machinations. To be so close to the stilled engine is like visiting an empty stage before a performance or walking down the middle of an avenue temporarily closed to cars. Steinbeck wrote about how "the sound of a jet, an engine warming up" could induce the "ancient shudder" of his wanderlust. Here, up close, is what makes that sound; what, exactly, warms up. Once the plane takes off, the space immediately behind the engine will be flooded with unimaginable speed and heat, easily 900 degrees Fahrenheit, spinning out as if from a ship's propeller into the icy vertical miles of nothing.

When I first looked closely at airliners, I was struck to see that national flags sometimes appear in mirror image on them—so that on the right side of an airliner the block of stars on an American flag is in the top-right rather than the top-left corner, on an Australian flag the stars of the Southern Cross shine on the left rather than on the right, and on a Singaporean flag the crescent moon appears to be waxing rather than waning. The idea, which appears in other contexts, such as the shoulder patches of soldiers, is that this is how even an image of a flag should fly as a vessel sails forth.

The sight of such flags is a reminder of the many ways an airliner answers to the air even before it moves. Often on a parked jet you find the blades of the engines are already rotating. Strong but light, built for air and the smoothest turning, they catch the slightest breeze and spin with the insistent nonchalance of a lawn mower. The things we have made, our air wheels that rest most easily in motion.

From inside the terminal on a windy day, the depowered rudders—the vertical panels at the back of the tail—of a row of parked planes may all be hanging to one side, blown all in the same direction like the branches of a line of windswept trees. The tail, which looks like a sail, acts like one, too. To counteract a crosswind blowing from left to right as we accelerate for takeoff, we must steer not toward the left, as you might think, but toward the right; the wind catches on the vast tail and rotates the nose *into* the wind, a phenomenon known as *weathercocking*. If you board a plane on a breezy day, you may feel it gently rocking back and forth before you leave the gate. That is mostly the tail, catching the wind.

Planes must occasionally be weighed, to ensure that calculations of the power required for takeoff, for example, are correct. This weigh-in takes place in a hangar, the doors of which must be kept closed, because even a light afternoon breeze on the wing will cause it to work a little, to soar ever so slightly, and tug the craft away from the scales of the earth.

It's October 2007. About a month ago I flew my last flight on the Airbus. We pushed back from Newcastle at 09:22, and parked at Heathrow just under an hour later at 10:21. Below that line in my logbook I switched to a different plane and a different color of ink. In the weeks since that flight I started my 747 type rating and now I've completed the classroom training portion and various exams. Today I'm entering the cockpit of a 747 for the first time. But there's only a cockpit. There's no plane attached to it. It is a box, surrounded by banks of screens, perched on jacks in a cavernous room. This is a full-motion flight simulator. Today I start the simulator sessions; I'm virtually flying. My first flight on the real airplane is already scheduled, for about a month from now: London to Hong Kong.

Though the simulator's wraparound video screens do an admirable job of conjuring up the cockpit's expansive views of the world, the simulator must also simulate the blindnesses that are a striking feature of flying a large airliner. The plane is so high, and the nose so rounded, that we cannot see anywhere underneath or immediately ahead of the plane. When we are taxiing, the knowledge that there is nothing under the nose of the plane in a given area comes solely through having seen that area before.

From the cockpit we cannot see anything behind us. When giving taxi instructions, air-traffic controllers must take account of this. For example, they may ask us to inch forward to allow an aircraft behind us to make a turn, the kind of maneuver that might occur naturally between courteous drivers with their rearview mirrors. On some airplanes the pilots can see nothing of the wings. From my seat on the 747, I can see only one of the four engines and a small portion of one wing, and even these only with difficulty.

Nor can we see the wheels, some of which are 30 yards or so behind us, or the tail, a further 40 yards or so behind them. This unseen length, and the enormous wingspan, means that maneuvering on the ground can be more challenging than flying. It's like walking while carrying long planks of wood: you must adapt your sense of size and shape and consider beforehand how you will move and turn. Sometimes a controller asks us to report when we have left a runway, and we must remember that while we, in the cockpit, have left the runway area, nearly all of the rest of the plane behind us has not. In aircraft manuals, elaborate charts that recall da Vinci's *Vitruvian Man* show the angles and distances that the extremities of the plane sweep when the plane turns on the ground. A pleasing terminology accompanies these images of the plane's turning limbs: *tail radius* and *steering angle,* and the wing-tip that *swings the largest arc.*

As much as possible this physical reality must be mirrored by the simulator. A taxiway light ahead will come toward me, then disappear under the nose. A certain distance, a certain speed-dependent time later, the simulator will shudder as the virtual wheels hit the same virtual light.

Pilots may jokingly start a reference to a bygone era of avia-tion with the phrase: "Back when Pontius was a pilot . . ." When I hear this it is hard not to remember that the cockpit screens that condense information about the far-flung ends of those airplane-spanning wires are often called *synoptic displays,* or the *syn-optics,* a term that means "seeing-together" and that may remind us of *synopsis,* or the Synoptic Gospels, the accounts of Matthew, Mark, and Luke that represent three takes on roughly the same events. There is a synoptic for the landing gear, which displays the tire pressures, brake temperatures, and the positions of the gear doors, and another synoptic for the air-conditioning, and so forth, for each of the major categories of systems.

On an older aircraft like the 747, the computerized maps on our *navigation display* are surprising for how little they show. Typi-cally a pilot looks only at the waypoints that will be crossed in the next quarter of an hour to one hour of flight, and perhaps a few nearby airports and radio aids.

What is not displayed on the 747—what the cockpit comput-ers do not even know about—is nearly everything else that might appear on any other kind of map. The cockpit's computerized map does not show the locations or names of a single city, state, province, or country. It does not show anything so lowly as roads or railway lines. It does not specify whether the ground is covered by forest or desert, or even whether it is ground at all, rather than a lake or glacier. It does not know of rivers. Even mountains are generalized into unnamed blocks of forbidden air; not features of the earth so much as failures of the air; aberrations in the purity of sky. There's no concession to the general-interest user or the geography lover, no flair or artistry, no Elysium or sketched dragons lurking along the edge.

In the cockpits of these vessels that round the planet, there's no way to zoom out far enough to see the curve of the earth, an increasingly common feature on the moving maps viewable by passengers. Although once a colleague on the Airbus showed me a convoluted method to conjure a waypoint on the opposite side of the world. When I was first shown this trick, we were over northern Germany and our antipode was in the Pacific, somewhere southeast of New Zealand. Once such a waypoint is plotted, if we've done it right, it will appear on our screen, as if we are looking through the center of the plane's spherical idea of the earth. This mark of our opposite location then moves *up* the moving map, a strange sight on a screen where everything else affixed to the earth—airports, beacons, mountains, whatever's on this side of the sphere—moves down.

When we enter the simulator the motion is deactivated. Before the instructor activates it, we must fasten our seat belts, as we must on the real aircraft before it moves. In smooth conditions in the cruise the seat belt is worn as a heavy-duty but otherwise ordinary version of what passengers wear across their laps; for takeoff and landing, however, we are attached to the seat by five straps that meet at a starlike buckle in front of us. The instructor demonstrates the oxygen masks—the web of straps that expand and then close over your entire head, like some life-giving squid bred expressly for this purpose. On the 747—but not its simulator—there is also an overhead hatch. There are handles, *inertial reels,* that under your weight will extend slowly on a cable to lower you all the way down to the tarmac, as if from the roof of a three-story building.

In certain training exercises in the simulator that relate to the presence of smoke or fumes in the cockpit, we are taught to pull an overhead handle that vents the cockpit air directly to the outside atmosphere. Cigarette smoke, of course, was once a regular presence in the cockpit. As I settle into my seat I am surprised to see ashtrays in the frames of the huge side windows of even the simulator, unused now. I smoked for several years in my late teens and early twenties. Cigarettes weren't a pleasant thing on airplanes, but I confess to a flicker of envy that previous generations of pilots were permitted to smoke in flight, sitting before the broad cockpit windows of a 747, "a cigarette in one hand, a map of creation / in the other," as Philip Levine wrote in "The Poem of Flight."

A few years later when I'm flying a real 747, I flick open the ashtray, and in the bottom of it is a tiny Hello Kitty sticker. I don't know who put this here, or why; I have to assume it was on a trip back from Japan, and probably after smoking was banned, as the sticker appears unmarked by ashes. I fly four dozen different 747s, and I never remember which one Hello Kitty flies on; until well into the flight, invariably over some desolate landscape, I idly open the cover of the ashtray and discover our unlikely stowaway.

One surprise the cockpit offers is how ridiculously small the switches that control important appendages and systems are. I was intrigued when I first learned to fly airliners that, unlike in a car, there is no ignition key. The simulator, which requires a computer log-on, and the occasional complicated reboot, is harder to get running than the real thing. The engines themselves are started, usually two at a time, with a series of thumb-sized switches. There are not many switches in the world that have so outsized but precise an effect as these; such enormous yet carefully calibrated consequences.

The sense of small controls determining large events is yet more vivid when it comes to the autopilot and its related systems. On long trips the autopilot is engaged for much of the time. Toward the end of the flight, when it is time to turn the autopilot off, the plane comes alive again. One quick press of a small button on the control column and the autopilot system instantly and completely disconnects. The plane is free. To take hold of the control column at this point and to turn it steadily, to watch the horizon tilt like a two-dimensional game board, the great panel of the world lifting, banking in response to my hands, is a feeling like nothing else.

Often, though, when I take family, friends, or guests of my employer on visits to the flight simulator, they are more amazed by the automation systems than they are by the experiences of flying or landing. Without touching the control column, in three or four seconds I could direct the plane to the North Pole, and then reverse my instruction and aim instead for the South Pole.

In the vertical sense autopilots have a number of different functions or modes—*VNAV,* or *vertical navigation,* is the telltale portion of modes like *VNAV PTH, VNAV ALT, VNAV SPD.* But even in the lateral sense—the left-right direction in which the plane is pointing—autopilots have a surprisingly varied menu. Because the plane may roll left or right around its long, nose-to-tail axis to fulfill the instructions given to the autopilot, these functions are collectively called *roll modes*, a kind of turning that the Wright brothers borrowed from the birds—vultures, in particular—they so carefully watched. One of the simplest roll modes turns the plane onto a selected heading, which it will then maintain across the whole world until the pilot selects a different mode. The dial for selecting the heading is called the *heading selector*. Turn this dial, roughly the size of a dime, to turn a 747.

Certain kinds of clouds tend to produce turbulence. Pilots can try to avoid these clouds by looking out the window and flicking the heading dial a few degrees this way, a degree that way. To spin this dial and feel 380 tons of aircraft roll to one side or the other, to see the earth tilt in response, is a thrilling sensation. Sometimes, the required heading changes even before the first turn is completed, and we must turn the heading selector in the opposite direction. My index finger and thumb barely move but the dial spins and the whole world, simulated or otherwise, tilts and rolls, as the craft obligingly reverses itself.

It's early December, a month or so after my initial visit to the simulator, when as a pilot I first enter a real 747 "for the purpose of flight," as the legal definition of departure intones. Unlike for my flights in the simulator, this time I need luggage, a toothbrush. And instead of walking across a detachable bridge to the floating box on stilts in a cavernous warehouse, I walk through the terminal full of passengers, down the jetway, and onto the main deck of the craft that is already being loaded and victualed. I climb upstairs and walk down the upper deck toward my new office. After so many weeks in the simulator it comes as a surprise that the rest of the plane—all of the vessel that lies behind the cockpit, that the simulator so diligently imagined for us in the long hours we sat surrounded by its wizardry—is present.

We load the route to Hong Kong, which is four times as long as any route I have ever before loaded into a flight computer. On these very long flights—London to Hong Kong, a route that hasn't been possible nonstop for all that many years—the historical backlight to the string the digital waypoints form across the earth is unmistakable. The plane can't know that perhaps no other pair of cities is so tightly bound across such physical and cultural distance, though it might guess from its frequent journeys between the two.

At this hour we are going to use full power for takeoff, the captain reminds me. This, counterintuitively, is one form of noise-reduction measure. The extra thrust will help us reach the altitude at which we reduce power more quickly, and so the horizontal extent of our noise footprint is reduced. Full-power takeoffs are a rarity on the 747, which is blessed with more power than we'll almost ever need. Many pilots remark that the sound of its engines actually becomes lower pitched at higher thrust settings; a force that at its greatest magnitude is felt more than heard.

We lift off and speed through the night and then the 747, as aircraft on long easterly flights often do, devours the entire next day as well. So it's on the second evening of the flight that we bank around Lantau Island and line up for the northerly of the two runways at Chek Lap Kok, the airport, opened in 1998, that pilots still refer to as "the new one," if they ever flew the famously harrowing approaches to the old airport at Kai Tak. Skyscrapers and ferries flip past dimly in the haze below, followed by lights on the tops of hills, then more broad, inky water. A few thousand feet

above this new earth comes the electrifying moment when I press
the button by my right thumb to disconnect the autopilot, and the
button by my left thumb to disconnect the autothrottle, a sort of
cruise control that on the 747 is usually disengaged before landing. The humid dusk is depthless, there is no natural horizon; the
runway lights appear to float in this dark and featureless sky, then
gradually refine and clarify themselves.

Later that evening we are in a bar, where the senior captain
buys me a beer. He asks me how it feels, then laughs and waves his
hand in front of my face when I don't respond. I apologize and
snap back to Hong Kong, from my daydream about how we left
London so many miles and hours ago. I'd been thinking about
taxiing the actual airplane out to the runway last night, an experience I had already experienced so many times in the simulator;
and marveling at how two physically and visually identical experiences could be so different. I'd been hearing again the voice of the
controller who cleared us for takeoff. The captain acknowledged
this clearance on the radio in the standard way. He'd then held his
hands out, palms up, smiled—for he had known this day himself—
and said: "Mark, here you go. Let's go."

Air

Air, in general parlance, is almost always an absence. A synonym for nothing. "It came out of the blue." "It vanished into thin air." Air is so devoid of imaginative weight that Galileo's discovery of its actual weight shocked the scientific world. Imagine a cube of ordinary sea-level air, a yard on each side. It weighs about 2 pounds—about the same as a liter or quart of water. Or, the next time you are lying on a picnic blanket and you gaze up at the contrail of a passing plane, consider that a round column of air only half an inch across—about the diameter of your iris—running from the earth's surface all the way up to outer space, contains around 2.5 pounds of air. Or that a typical picnic blanket—6 feet by 9 feet, say—has around 50 tons of air resting upon it.

To me, the truth that air is as substantive as concrete remains as counterintuitive as any of science's most inscrutable revelations about particles that exist in two places at once, or the unseen dark matter we are told comprises most of the universe. Little in daily life suggests that the air weighs down upon me as matter-of-factly as water rests on the bottom of an aquarium; that each day I awake and stand up and walk through insensible thickness.

The writer David Foster Wallace once related a tale of an old fish who asks a pair of young fish how the water is that day. The young fish are mystified; they do not know what water is.

Flight reminds us to ask: How is the air? The air is so marvelously alive that when I land from a flight and later take a walk under an open sky, I know I have left all the most physically dramatic activities of the natural world behind me, above me. And when we fly we confront many of our earthbound assumptions that would otherwise remain as unchallenged as the apparent emptiness of the air itself. There is a reason airliners have devices called *air data computers*. A pilot must learn to speak of at least four kinds of speed, and several more of temperature, distance, and altitude. These aren't questions of terminology or curiosities that appear only at the extremes of experience. They are the truths of the medium that sustains us, truths revealed to us by flight.

Of course, those who love to fly might be grateful that it's so hard to consider the air directly. The imperceptibility of air is surely one of the reasons flying entrances us. We find peace in the watery flights of fish and the sun's rays through their clear medium; and many of us love to swim or dive for the same reason. When a bird or, behind closed eyelids, our dreaming self flies up into the sky, when something detaches from the low world, it appears to cast off not only gravity and place, but much of its corporality and boundedness. The grace of flight, of movement unsupported by water or wheels, by rock or grass or anything sensible at all, is the simplest explanation for its long reign as a synonym for transcendence.

I sometimes take the Long Island Rail Road between Jamaica Station near Kennedy Airport and Pennsylvania Station in Manhattan. On this journey I have one of those inconsequentially strict routines that guide so much of daily life. When I am going *to* the airport, I almost always walk a little further up on 7th Avenue before heading into the station, to have the most time in the fresh air, even if it is raining or snowing.

But when I come in to the city, I walk as long as possible inside the station, in case it is raining or snowing!—or to linger in the air-conditioning. To draw out my indoor moment I walk through the New Jersey Transit concourse, to exit at 31st Street. Here, occasionally, my eye is caught by a fragment of Walt Whitman on the station's tiled wall: "This is the common air that bathes the globe." (In another poem Whitman extolled the "fierce-throated beauty" of locomotives, "launch'd o'er the prairies wide, across the lakes, / To the free skies unpent and glad and strong"—reason to speculate on what he would have made of airliners.)

The substance of flight is that air appears to have none at all. But it's worthwhile, just occasionally, to try to consider Whitman's commonwealth of air as directly as we can.

You might think of a journey by air first as a length. Imagine the line of your flight, not along the earth as we normally envisage it, but above it in the sky, slicing through our common air. The path cuts a narrow, staggeringly long portion of the atmosphere, a nearly linear volume that you travel through, and breathe from, after the engines compress it into something thick enough to sustain you.

You change that common air. You leave footprints in it, in the form of the contrails, along of course with the other byproducts of combustion and, indeed, of your own respiration. The engines awaken eddies no less real than the ripples on a lake—occasionally seen directly in the rotors of the contrails, streaming like ballet ribbons—while the wings raise unknowable architectures of wind and leave them to drift through the heavens. When pilots encounter a bump we think is the wake of another, unseen aircraft, we may gesture out at the invisible air-realm and say: "Someone's been here before." (That someone may be us; when practicing turns in a small plane, you know you're rounding out a neat, more or less horizontal circle through the sky when you have the satisfaction of crossing the jolt of your own wake.) And the air you exhale on the plane, the plane exhales, too. Your breath trails across the world.

Alternatively you might imagine the air as a sphere. When I think of the sky this way I remember pictures of the earth taken from space. Imagine the glowing rim of blue, the air, as a kind of second sphere that surrounds its interior, its rock-and-water twin. The atmo*sphere*, something we might move through; something we can see. Something more than nothing. This vision of the air-sphere also suggests the truth that our air is shared, communal (the transcending speaker of Whitman's "Song of Myself," perhaps), that it's exchanged and balanced by the breaths of animals and plants, of volcanoes and the seas. This spherical conception of the air is a reminder that we live *on* the world, not in it. What we live *in* is the atmosphere: the glowing air-planet that envelops the rock and water.

Or you might think of air not as length or sphere but as depth. Here, again, there is truth and comfort in the natural analogy with water. Evangelista Torricelli, the inventor of the barometer, framed this in a 1644 letter: *"Noi viviamo sommersi nel fondo d'un pelago d'aria."* We live submerged at the bottom of an ocean of air. Ralph Waldo Emerson, too, would speak of our enveloping air-sea, a few centuries later, in "this ocean of air above . . . this tent of dropping clouds." There's a particular kind of airport weather report known as a *surface actual:* the latest dispatch from the surface of the earth, from the bottom of the air-ocean.

When you put your mouth over an empty plastic water bottle and inhale, the bottle collapses. Not, as we may think, because your inhalation pulls the sides of the bottle in, but because you remove the air that held the bottle's shape against the crush of the atmosphere. The two Sumo wrestlers of the atmosphere press equally on the inside and the outside of the bottle, and the bottle holds its shape, immobilized; when the inner one is removed the bottle falls in on itself. The bottle of water I open in an airplane at high altitude crinkles and compresses when I descend, as it would if I plunged it deep into the ocean. Under our aerial sea the heft of the sky presses down on us, as water does on deep-sea creatures, or on us when we dive too deeply. We walk along the bottom of our air-ocean, as unaware of our water as David Foster Wallace's young fish. Occasionally we board a plane and swim-fly upward.

The depth of the air: I have flown over London on the fifth of November, and I have flown across the United States, from end to end, sea to sea, on the Fourth of July. The height of fireworks above our backyards and barbecues is nothing, really, in the towering scheme of the atmosphere. Each perfect burst has the scale of a coin at the bottom of a pool, or a tiny fire-flower, as fireworks are known in Japanese, underfoot.

When we pass directly over a radio beacon at high altitude, the navigation computers show our distance from the beacon as zero. But we can also display the raw data from these receivers, and at the moment we consider ourselves to be *at* a beacon, the data may still show us 6 or 7 miles away from it. Away—as in above. Imagine sailing over the deepest ocean and passing above a watery light on the sea floor, its glimmer fading as it rises through the volume.

The typical altitude of an airliner is about the depth of the Challenger Deep, the deepest known point of the oceans.

I think of somewhere that is 7 miles from my home. It would take two hours' brisk walking, or seven minutes' driving at 60 mph, to cross the distance down to the surface of the planet. The air is as nothing, and yet there is so much of it.

We may also gain a sense of the air by considering the force it exerts. I turn up the music and extend my arm from the window of a fast car on a warm summer day. When I change the angle of my hand my entire arm jumps or dips excitedly, almost beyond resistance. My hand pushes down on the air—like a wing, like the skis of a waterskier racing over a lake. I turn my hand so my palm faces fully forward, and my arm snaps back, landing against the window frame, as if a waterski or an oar had turned inadvertently, to catch the full force of the racing water.

From the rear cabin, passengers, not pilots, have a clear view of the little panels—there may be several—on the back of the wing that swing up or down as the plane turns. These are *ailerons*— little wings. The aileron was the creation of the English classicist and inventor Matthew Boulton (his forward-looking 1868 patent was titled "Aërial Locomotion Etcetera"). Consider what happens when on one wing the aileron dips and on the other wing the aileron rises. It's not quite the full story, especially on modern fly-by-wire aircraft, but in the simplest terms we can imagine that on the wing where the aileron dips down, it deflects the air downward even more forcefully, effectively increasing the lift on that side of the aircraft. The whole wing rises in response. The opposite wing, where instead the aileron rises up, creates less lift, and so that wing lowers. One wing rises, the other falls. This is *banking* or *rolling*, part of how a plane turns.

Imagine a bike or car that you could easily steer with small alterations to the angle of your hand extended in the wind—by "warping the gale," by "blading the wind's flank," as Hart Crane ("What marathons new-set between the stars!") put it. This view from the passenger's window seat—of the wing and the world in their proper arrangement, and the whole sky vessel rolling in response to the gentlest stirring of small panels—is not available to the pilots of many large planes, though there is no better picture of the invisible air.

It is an irony of flight—particularly of modern high-altitude, high-speed flight—that while an airliner inhabits the air even more fully than a boat inhabits the water, it quickly leaves most of that air behind. It's true that the atmosphere is a desperately thin band of air around the earth—thinner, comparatively, than the skin of an apple, as warnings about air pollution often remind us. But the greater shock is how quickly the atmosphere attenuates, even within this thin skin. Air is not distributed evenly throughout the atmosphere. Unlike water, which is barely compressible, the air accumulates at the bottom of its ocean; it piles up under its own considerable weight.

If I picture the earth from space again, I might remember that it is about 8,000 miles in diameter. To climb 3.5 miles—about half-way up to a typical airliner's cruising altitude—is to have barely left the earth at all. Yet at this altitude about half the earth's air is already below. You could still breathe here though, albeit with difficulty—most people climb Mount Kilimanjaro, which is nearly 4 miles high, without using oxygen tanks. Higher up still, at a typical cruising altitude of 7 miles, around four-fifths of the world's air is already below the airplane. Airliners do not yet enter space. But in terms of escaping the atmosphere that sustains everything we know, they get us most of the way.

Such a rapid fading of the atmosphere is related to one of aviation's more curious revelations about the air: that altitude itself is a fluid concept.

Even before a plane takes off, there is nothing simple about altitude. The earth is not a perfect sphere. Its true, misshapen character must be approximated for purposes of navigation, and whichever magisterially monikered model is used—World Geodetic System 1984, for example—must be carefully noted on our charts. Altitudes may be referenced to *mean sea level*, but this, too, is only an approximation; sea level depends on the tide, on the season, on which side of the Panama Canal you are on.

The elevations of airports vary hugely. Mexico City's airport, for example, is around 7,300 feet above sea level; Amsterdam's Schiphol stands below sea level. Pilots may joke that a flight from Mexico City to London is downhill, and so it is. Even airports themselves, even individual runways, which we imagine to be necessarily flat places, are markedly three-dimensional. In Dallas, the official airport elevation is 607 feet, but the elevation of one runway's threshold is nearly 100 feet less. In Mongolia, at Ulan Bator's airport, the elevation difference between the opposite ends of the same runway is more than 200 feet. That is nearly the height of a twenty-story building, or a 747 standing on its tail.

As creatures of the air, it's fitting that planes calculate their altitude by measuring air pressure. The air lies most heavily on places that are lowest, the places that have the most air piled above them. A *barometric altimeter* equates high air pressure—lots of air weighing down—with low altitude. The air, we're told, is as weighty and real as books stacked on my outstretched hand. On the ground the altimeter feels the weight, we might say, of ten books, and converts this reading to an altitude. Then, as a plane climbs, there is less air above it, and fewer books. The altimeter senses less air weight, less air pressure, and reports a higher altitude.

But there are problems with this simple equation. No device can be perfectly accurate, not least when it is mounted to the outside of a moving airplane. In addition to such *instrument errors*, another problem arises from how altimeters convert air pressure to altitude. They use a formula known as the *standard atmosphere*. It is an averaged model of the ideal sky, an air-Esperanto, a paradigm of how altitude, pressure, and temperature typically relate to each other on our home planet. But the actual conditions on any given day will never exactly match the standard atmosphere.

These errors may be quite large. Imagine you are on top of a mountain in autumn. An altimeter registers the weight of air pressing down upon it and accurately equates this with an altitude of 10,000 feet. But on a winter's day, in the same situation, the cold will cause the air to become denser and sink. More of the atmosphere's air will pile up below the mountain, and less will be above the altimeter to weigh down on it. This decreases the pressure around the altimeter. All the altimeter knows is that there is less air-weight, and so it reports an altitude higher than its true altitude.

So even in sophisticated airliners, pilots must manually apply *cold-weather corrections* to the heights of mountains. We might say that we'd normally wish to pass above a certain 10,000-foot mountain at 12,000 feet; but it is so cold out, we decide to treat the mountain as if it were 12,000 feet, and overfly it at 14,000 feet instead. This gives the all but inescapable impression that granite grows in winter—that when cold falls over the land, mountains rise further into the sky, until spring comes and they descend.

A more fundamental problem for altimeters is that air pressure varies in each individual place as both time and weather pass. It also varies between different places on the earth. Imagine an airplane parked on the ground as a low-pressure system arrives over an airport (generally, high pressure is associated with fine weather, low pressure with poor). The altimeters sense less air weighing down from above. Who knows if a lower-pressure system has arrived, or if the plane has started to climb up in the sky? Not the altimeter.

Similarly, a high-pressure system moves in, and there is more weight of air on the altimeter. Is it weather, or has the plane descended? The altimeters can't tell. It's a regular occurrence to board an airplane that has spent the night at an airport at sea level, say, and for all the altimeters to claim that I am well underground, or already, as far as the altimeter knows, flying slowly upward into the sky. If the pressure is changing quickly, then even though we are parked at the gate, utterly stationary both vertically and horizontally, our displayed altitude will rise or fall before our eyes.

The anomalies caused by such variations are dealt with by adjusting the altimeter to the local weather conditions; we give it a new starting point so that it can calculate our altitude from the hour's pressure more accurately. One of the first tasks to be carried out when we sit down in the cockpit is to obtain this *pressure setting*. As we dial it in, the altimeters spin contentedly in response, and the plane, for all it knows, rises from beneath the earth or descends toward it. In making these adjustments in the first minutes in the cockpit I feel a mild sense of mental exhalation, a relief that at last the air and ground of a morning are properly arranged.

Controllers check that we have the correct, most current pressure setting. Not only is an accurate pressure setting crucial to each pilot's awareness of their height above the ground, but the safe vertical separation of aircraft from each other depends on all the planes in a given area referencing their altitude to the same pressure setting—the same model of a given hour's air. When the air pressure changes, air controllers may announce the change over the radio—to "all stations," i.e., all aircraft. When such a pressure change is given all nearby pilots dutifully adjust their altimeters. The displayed altitudes in all the planes approaching Heathrow or Atlanta or Dubai then all change at once, harmoniously flickering upward or downward together. These pressure settings are so important that we have formal procedures to ensure the multiple altimeters on an aircraft are all set correctly. If they are not set to the same pressure the plane may flash a message to us: "BARO DISAGREE" (*baros*, meaning weight—of the air).

In my first year of commercial flying at least a dozen captains recommended the 1961 book *Fate Is the Hunter* and two gave me copies of it. Written by the aviator Ernest Gann, it contains many eye-watering tales from a former age of aviation. In one story, as their ice-shrouded plane descends toward mountains, Gann asks his captain, as if it is an ordinary question, whether it's now the right time to purchase some altitude by throwing the passengers' luggage overboard. Later Gann describes a harrowing descent in fog over water. He's trying to come down sufficiently low to find Iceland, where he has no choice but to land, without going so low that he hits the ocean, which he is unable to see. His air pressure–based altimeters are functioning normally but, because he does not have a local pressure setting, he has no idea how high he actually is. In the end a colleague dangles a cable from the back of the plane and waits to feel it snag on the churning surface of the North Atlantic. "When you feel a tug," Gann orders his colleague, "yell your head off."

Such stories make clear the importance of adjusting altimeters to local pressure when you are flying near the ground (or water). But even more surprising, perhaps, is that in the high, long hours of flight between airports, airline pilots abandon these local corrections entirely. We switch the altimeters to *standard*, a pressure setting derived from the standard atmosphere, that universal model of the earth's air. In doing so, we shrug off the actual weather of the day—the hour-to-hour, place-to-place vagaries of the real-world atmosphere.

To ignore local air pressure, of course, is to ignore our true altitude. And this is just the collective inaccuracy that high-flying airliners embrace. Whatever the altitude shown on the screens in front of you in the passenger cabin, whatever is displayed on the altimeters in the cockpit, the plane is almost certainly *not* at that altitude, because the pressure of air on the earth immediately below you is not known and, even if it were—from the weather report of a nearby airport, for example—our altimeters are not set to it.

Even more curious is that airplanes following an altitude referenced to the standard atmosphere collectively and continuously adjust their measure of wrongness—gently climbing or descending as the true pressure around them changes with time, and as they move across the world into different realms of weather. It was a memorable moment of my training when I realized that a plane flying at 35,000 feet is unlikely to be at the same altitude as another plane, elsewhere in the world, whose altimeters also show it to be at 35,000 feet; or that if a plane could somehow hover in one place, precisely maintaining its standard idea of 35,000 feet, it would in fact slowly rise and descend with the weather.

You might think of an ocean, of all the boats across its vast expanse rising and descending on their local swells; the simultaneous localness and global interconnectedness provided by the water. All the boats are on the surface, though their true elevation varies. An altitude referenced to the standard atmosphere is like such a surface: a membrane of air, pressed with indentations and textured with rises, shimmering invisibly over the aerial imperfections of the world and the air that lies on it. The world's high-flying planes follow these both vertically and horizontally; from one moment and place, into other moments and other places.

The high altitudes displayed in the cockpit are thus so detached from true altitude that they are termed *flight levels*, not altitudes—a distinction lost to both the moving map in front of passengers and every altimeter in the cockpit. Flight levels, then, though invalid to the extent that we colloquially equate them with true altitudes, are just what we might expect of an industry that works on a single time zone. They are both a leveling and a fiction; a globalization of the sky. Such a system—though its embedded inaccuracies may be surprising, and though some newer airliners allow pilots to call up a display of GPS-derived altitude—is both safe and purposeful. Many aspects of an aircraft's performance are referenced to the standard atmosphere, and a shared, fixed altimeter setting ensures that nearby airplanes are properly separated from each other.

There is a final idea of altitude that airline pilots must learn—

radio altitude or *radar altitude*. Radio altimeters bounce a radio signal off the earth and calculate their height from the amount of time it takes for the signal to return. The radio altimeter only cares about how many feet of measurable space are directly below it—a figure that it often announces to us out loud in the cockpit. At low altitude their accuracy circumvents the vagaries of air pressure and the varying elevations of the hills surrounding airports, and in the vertical dimension, at least, radio altimeters partly replace the eyes of pilots during automatic landings. Radio altimeters are so precise that they must account for the time a signal takes to travel within the wiring on the aircraft itself. They are extremely reliable; though curiously their radio eyes can *lose lock*—lose traction, over certain kinds of ground cover such as blowing, long wet grass (a problem more for the pilots of helicopters than of 747s).

The radio altimeter is the most precise measurement we have of our distance from whatever is immediately below us. But its very precision creates further conundrums. Sometimes at high altitude it will catch its reflection not on the ground but on another airplane. Though our air pressure–based altimeters show we are at 38,000 feet, the radio altimeter may faithfully announce "ONE THOUSAND" in the cockpit when we overfly a plane at 37,000 feet. That's a number we expect the radio altimeter to register just before landing, when its interrogations are returned to it not by another airplane crossing under our radio shadow high over Mali or Missouri, but by the approaching earth.

The designers of radio altimeters must also consider the deceptively simple question of where, exactly, zero is. That is, where the airplane itself begins. Given the radio altimeter's use near landing, it makes sense to define a height of zero not by the bottom of the fuselage, where the radio altimeter itself is likely mounted, but by the bottoms of the wheels when the landing gear is extended. But this isn't straightforward either. When a 747 comes in to land, the landing-gear legs—shock absorbers, essentially—are longer because they are not compressed. Additionally, the plane itself is nose-high, tilting upward. (Many airliners point upward not just in the climb, but throughout the cruise and much of the final descent, too, an upcast geometry of flight that partly explains the brake pedals on meal trolleys, the subtleties of flat beds, and why it is almost always harder to walk toward the front than the back of a plane.)

Because in flight we want to know the height of the lowest point of the plane, the radio altimeter starts counting not from the altimeter itself, or from where the wheels end when they are on the ground, but from roughly where the wheels end when there is no weight on them, when they are flying freely through the air.

At landing, though, the nose lowers and the weight of the plane presses down onto the landing gear, compressing it. Now the 747's ever-truthful radio altimeter finds itself below where it understands the ground to be in flight. And this is exactly what it reports to us when we first walk onto a parked 747, adjust our seats, turn up the brightness on the screens, and start our takeoff preparations: that this airplane is 8 or 10 feet below the earth's surface. Even the most sophisticated measure of our height above the earth's surface is inconstant; it depends on whether we are coming or going.

One October a friend and I went to Iceland. We drove clockwise from Reykjavik, and late one night, several days and autumn storms later, we rounded the island's far southeast corner. In England the weather makes perfect sense to me if I imagine, even in the center of London, that I am on deck, at sea; if just past the newsstand or coffee shop on the far corner of the street I picture a pounding, Turner-caliber seascape. Driving in Iceland I repeatedly had the sense not that we were at sea—though the sea was almost always nearby—but that we were flying. I've never been more aware of the wind's effect on a car. Merely staying on the road required a near-permanent force to be applied to the steering wheel, left or right depending on the direction of the road and the rain-laced crosswinds. Each gust knocked us halfway out of our lane.

Planes, too, must sometimes be briefly driven down a road in strong winds, on the runway at takeoff or landing. A plane on the ground can be steered with either the wheels or the flying controls—or with both. As the speed of a plane on the ground increases and more air flows over its *flight controls,* these grow in effectiveness, as a hand-wing extended from the window of an accelerating car might. This gives the unexpected and accumulating sensation that during takeoff you are simultaneously driving in the air and flying along the ground. In Iceland, after we stopped one night, I thought how much easier the driving would have been if Icelandic rental cars were equipped with some airplane-style controls, a rudder, perhaps—some recognition of and accommodation to their unintended life in the air.

Wind, to the earthbound observer, suggests a local event against a default background of stillness, a breeze passing over a fixed point on earth; over us. But higher up, the air itself, the reference frame of flight, is almost always in motion. Once I leave the ground I no longer think of wind as a flow of air that passes over us; rather, it carries us whole, as a river or an ocean current.

If you could produce a view of the earth showing only those things moving faster than 100 mph—a map of speed, a worldview made only of motion—you would see a few trains, and plenty

of motorists in Germany, the lines of their velocity sketching out
the network of autobahns. You would see many planes, material-
izing as they accelerated for takeoff and vanishing from the speed
planet when they landed and slowed. Mostly, though, you would
see the *jet streams*, the high winds that ring the earth. "Where are
the jets tonight?" I might ask a colleague, or comment that: "All
the way over we've been fighting a strong jet." The *jet* in jet stream
is said to derive from the streaming quality of the winds, not the
aircraft we most associate with them. But these winds were only
properly understood after we started to fly, and today the name is
a pleasing convergence of the engine type and the wind patterns
that both enable and shape our greatest journeys over the planet.
The fastest jet stream I have flown in recently was 174 knots—a
tailwind, thankfully. (Knots are nautical miles per hour; a nautical
mile is equal to 1.15 regular, statute miles, so 174 knots is around
200 mph.)

There are many factors that determine which path a plane will
follow between two cities. Sectors of airspace may be congested or
temporarily closed, often due to military exercises, and the varying
navigation charges that countries impose mean that routes longer
in time or miles can nevertheless be more cost-effective. En-route
weather conditions are another consideration. But in the absence
of such factors, the primary task that flight planners and pilots
face is to navigate the high winds; to harness them by hitching
a lift on a sky river that is flowing the right way or to actively
avoid them, fleeing the tempests that would enormously slow an
airplane's progress over the world.

Over the North Atlantic, which so many planes cross en route
between North America and Europe, a new set of wind-optimized
routes is charted each day, one set for the westbound flights and

one for the eastbound. Each day the westbound planes may arc far to the north, swinging high onto the Labrador coast to avoid the west-to-east winds that blow further south. That night those same planes may return to Europe in a momentous arcing, a vast and more southerly swoop, seeking out the heart of the eastbound jet stream that only hours earlier they went so far out of their way to avoid. Often the winds lever the paths of opposite-directioned planes away from each other so effectively that the route of an airplane flying from London to Los Angeles, say, will never once cross the path taken the same day by a plane from Los Angeles to London.

A *great circle* is the so-called straight line between two places, as a string would connect them around the surface of a sphere. (Indeed, in my airline's office, to one side of a bank of flight-planning computers stands a globe with such a string still attached to it, a relic of the age when great circles might still be plotted by hand.)

On flights from northern Europe to western North America, passengers are occasionally surprised to see from the moving map screen how far north we are—often over Greenland, where one of the planet's most reliably spectacular panoramas of sea, ice, and mountains awaits those travelers lucky enough to have a window seat. The shape of these routes is often attributed to the marvel of great circles. But westbound planes often fly still further north than the great circle to escape the howling headwinds that would devour their time and fuel; the same winds for which that night's returning pilots will be grateful, when they fly well south of the great circle, seeking the sky where the eastbound air tide is strongest. Occasionally I've started a flight from London to the west coast of North America on airways heading slightly northeast around a congested bit of airspace, rather than northwest. The overriding task on such a day is to escape the westerly wind. Only then do we take up our course.

While planes may move to or from these rivers of air, the rivers themselves transform and wander over the earth. They strengthen or weaken, twisting and drifting away from their typical haunts in long, languid sweeps across the cold miles of sky. The optimal route—the shortest and most fuel efficient—between two cities changes constantly, and so the route that an airliner flies between them can vary dramatically from one day to the next. Some days when I fly to New York, the last I see of Europe is Northern Ireland; the next week, flying to the same city, the farewell takes place over Land's End. When the jet streams drift far from their traditional homes, skies that are normally quiet will fill with jets for a few hours or days, their pilots drawn like surfers to the most favorable sky-swells. The natural forces of the world shape even these, our most technologically advanced journeys; they guide our migrations with an all but biological simplicity.

Planes also alter their vertical paths in response to the wind. At each point in a flight, a plane has an optimum altitude, based primarily on the aircraft's weight. As fuel is burned the aircraft's weight reduces and the optimum altitude rises. So in an ideal sky—one free of other airplanes and variations in the wind—a plane might climb continuously throughout its flight, until it was time to start the descent for arrival. But vertical differences in the strength of winds often overwhelm this ideal altitude, and so we may climb up early, to an otherwise inefficient altitude, because it is even more valuable to us to find the heart of a jet stream. Or we may descend to avoid one. On a recent flight from London to Miami, the Atlantic headwinds were so strong and wide that there was no way to avoid them horizontally. So, after climbing to 37,000 feet for the first few hours of flight, we descended to 29,000, surrendering nearly a quarter of our initial altitude, more than enough to make our ears pop. Later we climbed again, and then we descended; a kind of porpoising, a vertical wayfinding in the ocean of air.

Our flight paperwork and the onboard computers help us make such calculations. But we also have paper tables in the cockpit called the *wind-altitude trades*. The name recalls the trade winds that swept ships to the New World, ships that would then arc northward to catch different winds and currents home, in an echo of the aeolian geometry that pilots and flight planners deploy every day above the same waters. Another pilot at a different altitude might ask us what wind we are experiencing—our *spot wind,* it's sometimes called—which will help them decide whether to climb, descend, or stay right where they are.

As with altitude, an airliner is also always calculating the most efficient speed for the moment-to-moment conditions of flight. You might think, as I did, that this most efficient speed would be the same regardless of whether the plane is experiencing a headwind, a tailwind, or no wind. But in the tangled calculus of the air, the damage wrought to an aircraft's efficiency by a headwind is greater than the gift the same wind speed would offer, were it a tailwind. The stronger a tailwind, the less time you benefit from it, while the greater a headwind, the longer you are subjected to it. For this reason, the flight computers will suggest accelerating to what would otherwise be a less efficient speed to minimize the time we're exposed to a headwind. While in a tailwind the computers will suggest slowing down; to linger awhile with the wind at our back.

Wind is so critical to flight calculations that some manuals refer to a new kind of distance: air miles, or *air distance*. Air distance adds the effect of the wind to the distance along the ground that the plane flies over. As a unit of length, it is as fluid as the air itself; the high-country mile. If there is any headwind or tailwind at all, and there almost always is, then a flight from London to Beijing, for example, will have a different air distance than a flight that follows precisely the same route in the opposite direction, from Beijing to London—and both these air distances will differ from the distance along the ground.

Occasionally, while still on the ground, after I load the route into the computers but before I enter the expected winds, the flight computers flash up a warning that the plane does not have enough fuel to complete the planned route without dipping into its sacrosanct final reserve of fuel. Then I tell the computers something about the winds over the world; the computers consider it and are satisfied. Though the miles on the ground are unchanged, though we have not even started to move, the air distance has been remade by the wind.

Richard Bach, the author of *Jonathan Livingston Seagull,* once titled an essay "I've never heard the wind." Few pilots today ever do. Though they make up one of the natural world's most dramatic physical presences and determine both the path and length of all our journeys, the winds exist largely as numbers on the cockpit's screens, where a diminutive symbol (only an arrow, sadly, not a cockerel) turns like a weathervane. The wind's dimensions are a regular topic of conversation among both pilots and flight attendants on a trip, as casually central to our daily life as delays on train lines or morning traffic jams. We do not mind a headwind on the outbound journey, because it often means a faster return home the next day. But we do not hear or feel the wind directly.

Jet streams occasionally produce turbulence, particularly at their boundaries. But often they are so smooth they suggest something like the opposite of their true dimensions. When it's a little bumpy I remember that I am flying at 600 mph, in air that is moving at 200 mph, yet the plane is steadier than a car on a dirt road. Often the fastest winds are as smooth as glass. When the computerized wind readout shows even a routine wind for an airliner to experience, 50 knots for example, I think that anywhere but Iceland, perhaps, such a wind on the ground—58 mph—would make the news. We would struggle to stand against it, and yell to be heard.

Maritime cultures, such as those around the Mediterranean, still deploy many archaic names for winds—the Bora, the Sirocco, the Khamsin. Today, England has one named wind, the Helm Wind, known for occasionally shrieking down the western slopes of the Pennines in Cumbria. But it's easy to imagine that the so-called Protestant wind that blew the Spanish Armada away from England might have become a general term for the east wind ("Popish" winds blew, too, a century later, to delay the arrival of William of Orange). America retains a few named winds, such as the Santa Anas of southern California and the Chinook, and even a fictitious wind, the Maria, from the Gold Rush musical *Paint Your Wagon* (from which the singer Mariah Carey gets her name and its pronunciation). Hawaii once had hundreds of named winds; whether you could list a place's winds and rains there was a test of whether you were truly a local.

Imagine looking up one morning and seeing a Nile or an Amazon in the sky, in a slightly darker blue, a shimmering, partly reflective navy hue, twisting and curling in the north sky over your hometown, and migrating to the southern sky by the time you go out for lunch. Aside from the sun itself, such air rivers would be the most dramatic feature of the earth or the sky. We have so matter-of-factly fashioned the souls of cities, of literatures, of whole civilizations from rivers—the Danube, the Mississippi, the Yangtze. One manufacturer has even named its jet engines for British rivers—the *Spey*, the *Trent*, the *Tay*—to contrast their smooth flow with the unevenness of piston engines.

We might have made something more of the jet streams, have worshipped and built a complete mythology around their image, had our prescientific eyes been able to see them. Flying, though one day it will feel as old as sailing does today, is still new to us. It's not nearly too late to style the sky's high winds, to scatter up the seeds of an aerial heritage.

Though the jet streams run in sweeping, largely east–west bands around the earth, we might choose to give their different sections different personalities, as in England the names of some rivers and streets may change midstream. The high winds over the continental United States might be renamed the Wiley Posts, or the Post Winds, for Wiley Post, the one-eyed American aviator who was the first pilot to fly solo around the world, and who is partially credited with the discovery of the jet streams. Post died in Alaska in 1935, aged only thirty-six, in the crash that also killed the humorist Will Rogers. There's an old-style aviation beacon named for Post and Rogers on top of the George Washington Bridge in New York, a memorial in light that once also served as a marker for aircraft taking up their westbound routes. "Our flight to Raleigh," a pilot returning from the west might say, "is just under four hours tonight, thanks to the Post Winds." Meanwhile, the winds over the

North Atlantic, which blow so reliably and steadily from North America toward Europe, we could dub the Allied Winds, to help us remember they were best documented during the transatlantic resupply efforts of the Second World War.

It is a joy of my job that I'm occasionally tasked with drawing the winds. In the cockpit of the 747 many of our paper maps have been replaced by electronic versions. But there remains a set of disposable maps that cover the world, which are sometimes called *progress charts*. As a child I saw carbon copies of these mounted in the passenger cabin of the aircraft—the path of the plane, a charcoal-gray zigzag over blue sea and yellow land. I still have one or two that I asked the cabin crew for on those long-ago flights.

These charts have space for the date and flight number, and for the pilots to write their names and ranks. Completing a progress chart, I feel as if in another age I would be crouched over a table, in heaving seas, an oil lamp or heavy brass navigational instrument on the table to hold the paper in place. I like to draw the steady lines between distant waypoints in thick green or blue ink, to sweep over countries, mountains, oceans as only a pen, or an airliner, can do. Although we have other, computer-generated maps that show the predominant winds along our route, some pilots still draw the winds onto the chart, along with the forecast areas of turbulence.

A pilot who does not wish to waste paper might save the chart for the flight home, plotting the return route and winds in different colors from the outbound ones. But if we ever have visitors to the cockpit after landing, especially kids, we will gladly give them the chart to take home. Someday even these last of our paper charts will be removed from cockpits; they will become relics of the early jet age. For now they remain, these maps of transience and air in every sense—hand-drawn charts of one day's unique and wind-sculpted journey and of the great unnamed rivers of the sky that hindered us or blessed us and carried us on our way.

If altitude and distance are not straightforward concepts for planes flying high, neither are temperature and speed. Generally speaking, the temperature of the air drops as you climb, in the same way that mountains are usually colder than lowlands. An airliner climbs to high lands indeed, to a bitter world where temperatures routinely drop to minus 70 degrees Fahrenheit.

Temperature affects many things on an airplane—the efficiency of the engines, and the formation of ice that can disrupt both the engines and the airflow over wings. Measuring temperature, however, isn't straightforward. In cold climates meteorologists warn about wind chill, about how the wind can make an icy day feel colder still. Airliners go so fast that they experience not only wind chill, but also what might be called wind heat. Fast air hits a thermometer and is brought to a sudden halt, compressing and warming greatly—an effect my brother is familiar with from pumping up the bike tires I'm always asking him to help me properly repair. This effect, part of what's called *ram rise,* means thermometers in the slipstream generally report a temperature that's much higher than the air's ambient temperature.

On the Concorde, ram rise could heat the nose and the leading edges of the wings to over 212 degrees Fahrenheit—hot enough to boil water at sea level, and to cause the plane to stretch by around 10 inches in flight. On a 747 moving at less than half the speed of a Concorde, the effect is more modest. But a thermometer in the slipstream often still reads substantially higher than the actual temperature of the air—often at least 50 degrees Fahrenheit higher.

The temperature that airliners experience at speed is therefore called *total air temperature,* or *TAT* (rhymes with "hat"). It is distinguished from *static air temperature,* or *SAT,* which is the temperature the air around the aircraft would be had it not been compressed.

It's reasonable to think of SAT as the real or actual temperature, and to find in the difference between SAT and TAT an inexact if pleasing comparison to a tenet of quantum physics, in which the act of measurement may distort or alter just what you are hoping to measure. The higher sensed temperature, though, is not a question of measurement. The forward-facing parts of the plane—such as the leading edges of the wings and the nose—are known, appropriately, as *stagnation points.* These entire surfaces experience the same heating effect as the thermometers do.

This heat, though problematic for the designers of supersonic aircraft, can be useful. Consider the fuel in the wings of an airliner. Fuel cools dramatically during a long flight in the high cold, but it cannot be allowed to cool too much. Typical freezing points of fuel are around minus 40 (a temperature that requires no C or F to follow it; it is the intersection of the Celsius and Fahrenheit scales) or colder. The static temperature of the air outside—the ambient temperature, shown perhaps on the moving map screen—is often colder than this. But the TAT, the experienced, wind-warmed temperature, is much higher. Indeed, nothing suggests the speed of airliners and the physicality of air quite like the fact that if the fuel starts to get too cold, the simplest way to warm it up again is to fly a little faster.

If flying overturns our everyday definitions of distance, altitude, and temperature, it scrambles our intuitive sense of speed most of all. In daily life we have only one idea of speed: how fast we move over the ground. If you ask pilots how fast their plane is going, they might pause before replying. They might say it depends.

In the sky there are four important concepts of speed. First is *indicated airspeed.* This is best imagined as the speed at which you'd guess you were traveling if you stuck your hand out the window and felt the air against it. In all but the most limited circumstances, indicated airspeed bears little resemblance to *true airspeed*—your actual speed relative to the air mass around the plane. Third is *ground speed,* your speed over the earth, which is perhaps nearest to our terrestrial understanding of motion, though it is irrelevant to everything about an airplane that has to do with the air, and often differs from indicated and true speed by hundreds of miles per hour. Finally there is *Mach,* the true airspeed of the plane relative to the local speed of sound.

The relationship of these speeds to one another is so peculiar that each is best regarded as a separate idea of motion. In the cockpit the four speeds are displayed in different places or at different times. The computers may automatically change the font size of one kind of displayed speed as it becomes less useful, or change what kind of speed a certain switch alters, or even seamlessly replace a display of one kind of speed with another that has become more relevant to a new stage of flight.

Mathematicians occasionally question whether the achievements of their field are created or discovered; I ask myself this about airspeed. I wish I could remember, before I learned to fly, what I might have guessed indicated airspeed actually meant. Why, I might have asked myself, do we need to specify *air*speed? Isn't speed enough? And what about that *indicated,* which makes the term sound both fuzzy and very precisely qualified, as if the quantity in question isn't quite real, or as if speed itself is the wrong word for what keeps us in the air?

The difference between *indicated* and *true* airspeed is something I might consider when I hold my hand out of a car window. Using straightforward numbers purely for illustration, at sea level on a windless day, driving at 50 mph, I might feel a certain force of the air on my hand. But now imagine I am on a road near the top of a high mountain, where the air is thinner. Though I am still driving at 50 mph, my hand would register less of a force from the slipstream because fewer molecules are hitting it. I am still moving at 50 mph, but it might feel like only 40 mph to my hand. We might say that my hand's true airspeed is 50 mph, but its indicated airspeed is now only 40 mph.

The airspeed indicators—the speedometers—on a plane act something like an extended hand. They stick out of the sides of the plane, into the slipstream. They measure the pressure of the molecules hitting them. From this they subtract the pressure of the air that isn't moving—that is, the background weight of the air, the weight that Galileo discovered air possesses (the same background measurement of pressure, by the way, that the altimeters use). Indicated airspeed can be thought of as nothing more than the pressure that speed adds to the air; the rough difference between what your hand experiences inside and outside the window of a moving car.

Indicated airspeed, then, isn't speed in any normal, earth-referenced sense. Rather it is the feel of speed, the feel of motion through air. *Air force* or *air feel* would be better terms for indicated airspeed. Indeed, on some vintage aircraft, the airspeed indicator is outside the cockpit—a little panel that is deflected by the slipstream and read off against a scale of numbers beneath it, hardly more complicated than those light-hearted weather stations that feature a rock hanging from a board and instructions like: "If the rock is swinging, the wind is blowing."

Airspeed, though, is not less useful because it is so inchoate and capricious. On the contrary, it is exactly what pilots need, because it is indicated airspeed, not true airspeed, that determines how much lift the wing creates. For this reason there are multiple, redundant systems to sense airspeed, and one of the most important checklists on the airplane relates to failures of airspeed indicators or, perhaps worse, a disagreement between them. It also explains why, when landing on a blustery day, you may hear the engines power up and down so often. If a sudden gust of headwind hits your car while driving, your hand out the window feels this sudden gust as an *increase* in indicated airspeed. In an airplane at just such gusty moments the telltale airspeed indicator spikes upward. In response the pilots may then reduce thrust to bring the amount of air feel back down to the target—until the wind drops, the speed slows, and power must be added again.

When we fly we leave behind many earthbound ideas, so it's appropriate that the difference between indicated and true airspeed, small at low altitudes, grows substantially as a plane climbs. To conjure the same lift as at lower altitudes a high plane must fly faster in a true sense, to ensure that the same heft of air passes over the wing. At high altitude a jet's true speed may be 500 knots—nearly twice the 270 knots reported by our airspeed indicators. A plane climbing at a constant indicated airspeed is in fact continuously accelerating, a sky-sorcery that will be undone only in the descent.

Ground speed, meanwhile, adds the enormous effects of wind to our true speed through the air. It accounts not only for our motion through the air but the motion of the air itself over the earth. As its name suggests, it's the closest to our traditional understanding of speed, which is why it is the speed typically shown on moving map displays. Imagine two boats traveling in opposite directions along a fast river. Their speeds over the surface of the water—their true speeds—are the same. But their speeds measured over the riverbed, or along the riverbank, are different. The boat heading upstream, fighting the current, is moving more slowly relative to the riverbank, while the downstream boat, carried by the current, is racing along the bank.

In the air, two planes traveling in opposite directions at the same altitude, one flying in the direction of a strong jet stream and the other flying against it, may have identical indicated airspeeds—the same feel of the air, the same number of molecules hitting their wings and speedometers. They may have identical true airspeeds, because their speed through the surrounding air—over the moving river—is the same. Yet their ground speeds may differ by 300 mph or more, because the surrounding air is carrying one quickly over the ground, while slowing the progress of the other. In a strong tailwind the ground speed of a jet can even exceed the still-air speed of sound, but the plane itself, in the frame of reference of the jet stream that carries it and surrounds it entirely, is not traveling supersonically.

Ground speed, though irrelevant to flight, is of great importance when getting into the air in the first place. At takeoff, though indicated airspeed determines when a plane can fly, ground speed determines when it will run out of runway. Air is thinner at higher elevations and temperatures, so a plane must use more runway length, or additional engine thrust for faster acceleration, or both, in order to gather the required feel of air over the wings. A plane that takes off at an indicated airspeed of 170 knots in lofty Denver or blazing-hot Riyadh is going much faster—is using up the runway much more quickly—than a plane taking off at the exact same indicated airspeed in sea-level Boston. That's one reason long-haul flights from the Middle East traditionally depart late at night, when the air is cooler.

Wind adds another complication to takeoff or landing. Once we are detached from the ground, the flowing wind forms its own frame of reference, which is why the air around a hot-air balloon is wonderfully quiet and still even when the balloon is traveling on a very steady breeze. Horizontally, such a balloon has an indicated airspeed of zero, and a true speed of zero, but a ground speed equal to that of the wind. Similarly, a plane in flight moves in the reference frame of the moving wind. But on the ground the wind passes over an airplane as it might over the branches of a tree, or a balloon that is tied to the ground.

Indeed, the airspeed indicators on a plane cannot distinguish between airspeed and wind, because to the probes, as to the wings, there is no difference. Taxiing around an airport on a breezy day, the airspeed indicators will flicker to life when you turn directly into the wind and drop off when you turn a corner and face away from it. They may register airspeed even when the plane's wheels are stationary, just as you might by extending your hand from a stopped car on a windy day. Such a stopped plane, in an aerial sense, is already moving.

This reverses the everyday idea that tailwinds are advantageous. A tailwind is indeed a gift to a plane—but only once it is well away from the runway; when the wind carries it like a balloon and the plane's speed through the air is added to the speed of the air over the ground. Before we are airborne the last thing we want is the wind at our back. If a plane is rolling down the runway at

10 knots in a tailwind of 10 knots, its airspeed is zero; and yet it is already devouring the available runway. A headwind, meanwhile, is a blessing. A plane parked in a headwind of 10 knots is already part of the way toward getting airborne, though it has not yet moved.

Our love of headwinds at takeoff applies equally to landing, when a tailwind is an unwanted addition to ground speed, to the rate at which we consume the runway. The desirability of head-winds explains why aircraft carriers will turn into the wind or accel-erate; for both takeoffs and landings they are seeking a headwind or making their own. Back on shore, of course, a typical runway can be used in both directions, and many airports have multiple runways to cater for the vagaries of the wind's direction. When the wind changes markedly, both arrivals and departures may be briefly delayed while air-traffic controllers reverse the flows of traf-fic coming to the airport and leaving it. This is how the unseen motions of air determine how you will arrive in a place; how the wind gods choose what to show you of a city, when you first come to it from the sky.

There is one further idea of speed that pilots must consider. *Mach*, a word that still sounds as futuristic to me as it did the day I first heard it, is the aircraft's speed—its true speed through the air mass, not the indicated airspeed that the racing of the jet through the thin air sums to—expressed as a portion of the speed of sound in the local air. Mach is a peculiar kind of speed—a ratio, and so without dimensions or units.

When I was first taught about the speed of sound, in school I suppose, I assumed that it was only of interest to humans because we use sound as a means of communication, or because it explained why the light from a distant bolt of lightning arrives long before the noise. I've since realized, however, that there's a good reason planes pace themselves against the same phenomenon that brings to our ears everything from Beethoven to thunder. There's nothing arbitrary about our focus on the speed of sound. There is a speed of sound in iron, rubber, and wood, for example (all of which are faster than its speed in air). What we call *sound* is a kind of wave propagating to us, whether the voice of an opera singer, the pitter-patter of rain, or the noise from jet engines that Joni Mitchell ("I dreamed of 747s . . .") called "a song so wild and blue."

The bow wave that develops when a boat moves faster than the water can move away from it is analogous to the shock wave produced by a supersonic aircraft. A bird floating on the water would not sense the waves of such an approaching boat, just as a bird in flight would not hear the approach of a supersonic aircraft. The resulting pressure buildup, this aerial bottleneck, is what we hear on the ground as a sonic boom. We fly on Mach in order to pace ourselves properly against the limits imposed by the speed of sound, an elemental quality of the air we live in.

Airliners—since the retirement of the Concorde—fly below Mach 1, the speed of sound. But at high subsonic speeds the air over the top of the wings can reach or exceed Mach 1. This can result in the formation of shock waves that destabilize the pressure distribution around the aircraft, before the aircraft itself has exceeded the speed of sound. In order to stay within the aerodynamic design limitations that this imposes, airliners are typically engineered to cruise between Mach .78 and Mach .86. A normal cruising speed for the 747 is Mach .85, 85 percent of the speed of sound, read out as "decimal eight-five," or "point eight-five," or "Mach eight-five," or just "eight-five." If we are catching up to the plane ahead, a controller might tell us we are *in trail* of a slower jet; they might ask us to "reduce to eight-four."

At low speeds on the 747, Mach is not even displayed. But at higher speeds—in what pilots may loosely term the *Mach regime* or the *Mach realm*—it is by far our most important measure of velocity. Fittingly, as we accelerate in the climb, our Mach number automatically appears in the same place on our screen where, when we were lower and slower, our ground speed was displayed. As we slow in the descent we switch again, from Mach back to airspeed. An air-traffic controller trying to space out aircraft descending through this boundary must therefore assign us two different kinds of speed to cover both the lower and higher speed dominions of the sky. "Start the descent at Mach eight-two," a controller might say. "Then on *conversion* [or on *transition*] fly 275 knots."

A further curiosity is that Mach, described to us by the science of fluid dynamics, is itself fluid. The speed of sound varies with temperature. So just as with altitude or indicated airspeed, the same Mach number relates to different speeds at different times and places. Mach is valuable not despite this variability but because of it. A plane traveling at a fixed Mach number will move slower when the air around it is cold and faster when the air is warmer. But at a constant Mach number, the aerodynamic conditions will remain the same. Sound, in other words, is as fluid as our most tender notions of music would suggest, and all our journeys in the higher sky are tuned to it.

Water

I'm in the window seat on the left-hand side of a sky-blue 747. I'm traveling to Belgium to spend the summer with relatives, but I'm flying first to Amsterdam, to stay for a few days with a family friend. I'm fourteen. It's the first time I've ever been on a plane without my parents.

I will come to think of the family friend I am visiting as my oldest friend, in both senses. I remember, even as a small child, thinking that she was not really a grown-up; that she was my friend as much as my parents'. But she is more important to me than that, even. It's in large part due to her that my parents met. A born-and-bred New Englander, she and her husband spent a year in the late 1960s studying poverty in Salvador, Brazil, where they met my father. My dad already had the idea that America might be the next place for him; maybe even the last place. His new American friends helped him make that decision. They were perhaps the only Americans he knew well—and the reason for his coming to Boston when he left Brazil. My mother met him at a talk he gave in Roxbury about his work in Brazil, on the weekend after the assassination of Martin Luther King Jr.

Two decades or so later, I'm flying alone to the Netherlands,

where my oldest friend has moved. Last night my parents drove
me from Massachusetts to Kennedy Airport in New York. Before
we left they took a photo of me in our driveway, standing in front
of our green Toyota, clutching my passport, which I now finger
in my backpack as we start our descent. We're only half an hour
from Amsterdam. My friend will have parked at Schiphol, she will
already be in the arrivals hall; she knows that this is my first flight
alone.

I'm listening to my new Walkman as I look out of the window.
It's the first time in my life that I listen to music on an airplane,
and for many years afterward this soundtracking, the overlaying of
music on the turning world, will accompany every one of my most
treasured experiences of flight, particularly at takeoff and landing. I will learn to pause or rewind, to pair the size of the growing
trees with the remaining minutes, to ensure that whatever song
I've saved for this moment is playing, or best of all just ending, as
the wheels reach the runway.

This morning I am lucky to have a window seat to accompany
my music, but unlucky, I feel, with the weather. When the sun
came up an hour or so ago the floor of the world far below was
white and without texture. Now the jet is descending into this solid
deck of cloud. The blue disappears, and then there is neither sky
nor earth in the window, nothing to suggest the distance nearly
completed between New and Old Amsterdam or anything at all
about the place I am coming to. There is only the noise of the
plane in the mist, and the occasional small jolt of physical sensation, to remind me that the structure in which I have spent the
night is in motion.

The tops of the high clouds that the plane first clipped and
then entered were brilliant white. Now we are gliding through

darkening gray, a detuning of day that's perfectly synchronized with our continuing descent; the altimetry of light. I remember what a science teacher once told me, that the clouds are not as we think. They are not water in its gaseous state. Water as a gas—the humidity in the air—is invisible. What composes the clouds is ice or small droplets of liquid water. Vapor or steam rises invisibly from a cup of tea. Only when it cools again to liquid droplets do we see the cloud that has formed in the kitchen.

A ship appears at some indeterminable distance below. I blink. I don't understand, because the ship appears to move directly up the windowpane, as if it is sailing vertically through the cloud. A moment later I realize the ship moves this way because the 747 I'm on is banking sharply over the surface of an ocean I cannot yet see. The cloud momentarily thickens and I lose the ship. Then we descend further; we are over a white-capped, churning gray water. I can see the line along which the cloud-sea we have descended from meets the water-sea below; the Dutch coast. From above the clouds, and within them, we might have been over any place, in any age. But below them it is this day, off this country, and I see that the plane I'm on is just one of many ships coming to the Netherlands from far across a generality of water.

Flying gives us what's perhaps the last thing an aspiring pilot would expect: a close experience of water. We think of the conceptual partition of water and air as elemental, as simple as the horizon. But airline pilots see more water than any sailor does. About two-thirds of the world's surface is ocean; much of the land that long-haul pilots work above is covered in snow or ice. At any given time about two-thirds of the world is covered in cloud. It is an extremely rare moment in a plane when you cannot see water.

Gray waters off Gothenburg lie under close pages of fog, early morning clouds perfectly map in water the roiling seascape of Scottish hills they cling to, subtropical Bahamian seas shimmer in their zoomed-in, blue-boundaried rainbows. Many Arctic lands hide under snow so all-encompassing that their surface is indistinguishable from a solid cloudscape or an ice-locked sea. For many miles and hours in the sky—sometimes for nearly an entire flight—water is the only thing we see.

Within the range of temperatures found on the earth's surface, only water exists naturally in its solid, gas, and liquid states. Together these three states compose what scientists call the *hydrosphere*. Seen from the sky, the hydrosphere, our round world of water, turns as guilelessly as a wheel, forming a cycle that could hardly be more archetypal. Rumi wrote: "I want your sun to reach my raindrops, so your heat can raise my soul upward like a cloud." On average a molecule of water spends as long in that sky as you, having flown through it, might spend on vacation: nine days.

One of the best reasons to become a pilot, especially if you are from a cold and often cloudy place, is the chance to surface from the world of clouds; to know that sunlight will be present on nearly every day of your working life. An overcast sky now appears different to me on the mornings of the days I am going to fly, because I know I will soon be on the other side, that the clouds, a backdrop of one low scene, are only a curtain drawn over a brighter and more elementary one. Above every gray winter day the cities of cloud are tumbling, rising, migrating, and dying in the torch of the sun. A free world of light and water self-shaping in the most liberated ideas of form, among which we pass our most ordinary hours on an airliner.

A forest or grassland below may reflect 20 percent of the sunlight that falls on it. Some clouds may reflect 90 percent. Typically, it is only when the world below turns from clear land to cloud, or we descend into the tops of such a sun-blasted cloudscape, or on the rare occasions when we are still in cloud at cruising altitude, that I put every one of the sun visors up around me, to curb the headache of so many large windows that are each as featurelessly and brightly white as the frosted glass of a fluorescent bulb. Or I put on my sunglasses, paired shields against the blinding majesty of sky water, which in the sky would be better termed cloud-glasses.

The deserts are so rarely covered in cloud that they predominate among the visible land on a long flight, which can give the impression that the lands of the earth are drier than they are. Then a city appears in such a desert and the water we see near it—lakes, dams, rivers locked in their rolling green frames of vegetation, twisting across desiccation—looks as holy as blood. We turn above the liquid life of the Tigris, the Ganges, and the Mississippi, the sun setting over the shining ribbon on the land, as civilizations flicker to life on the riverbanks like stars for the coming night. Here is the hydraulic shadow of civilization, the up-scattered light of water: Baghdad, Varanasi, Memphis.

In the DC-3, the 747 of a previous generation, pilots would sometimes wear rain slickers or boots in the cockpit, so low in the sky did they fly, with leaking windows. Now we fly above most of the low world's weather, which is one reason why flying is generally much smoother than it was in the early days of aviation. Most, but not all, and so our weather radars scan the path ahead. The radar's gaze pierces clouds and *returns* a map of precipitation— of larger droplets, closer-knit agglomerations of sky water—that is then displayed on the same computer screen as our route. A storm rising from the earth appears onscreen as overlaying, fractal pools of color-coded severity, red encased in amber, and amber embedded in green. Such a roughly horizontal slice of a storm is displayed directly over the clean line of the aircraft route and the icons of beacons, forming as well-matched a composition of the organic and the technological as a close-up image of a bacterium that includes the detailed tip of a scientific instrument.

We fly far around storms but at night, even at a great distance, their flashes may still fill the cockpit; we may throw the *storm* switch, which automatically sets nearly all the lights in the cockpit to their maximum brightness, so that distant night lightning will not blind us. The map of so many of my most-often-traveled routes is written in water or its absence; gray clouds over Europe, the clear deep volume of the Saharan dusk; storms strobing in the conurbations of cloud over West Africa, dawn over the flaxen desiccation of the Kalahari.

In daylight, seen with our own eyes, the rain streaming from a distinct cloud resembles nothing so much as beams of light. It is routine from the cockpit to see the storms rising and clouds forming, blistering upward or vanishing in real time, and to see from them the fall of new rain on the roof of the ocean, or to overfly the endpoints of glaciers, where shards of the ancient snow glass shatter in the sun and tumble into the police-light blue of northern seas. Often I look down on a sea that is streaked with white, and I cannot tell if these apostrophes of white-turned water are the work of the wind-whip cracking on the far-below waves, or if these whitecaps are in fact clouds of ice.

At most latitudes, the sky over the ocean is more likely to be cloudy than that over land. But even over the ocean a solid floor of cloud

can end suddenly. When we cross the coast of such a great country of cloud, we emerge between the earth's blue mirrors. Sunlight scatters through the molecules of the air; it falls onto the sea
and tumbles among the molecules of water. The results are the
very model of blue, the color, somehow, of both above and below,
both freedom and meditation: the "wild blue," the "deep blue,"
the "aching blue." From our aero*nautical* vessels the colors of the
ocean and the sky are often so perfectly matched that it is hard
to say, without reference to the horizon, which is water and which
is sky.

Robert Frost visited the coast of North Carolina as a teenager,
about a decade before the aerial labors of the Wright brothers
would come to fruition there. Later, Frost referred to his time on
that Carolina shore, and what took flight from it, in a poem titled
"Kitty Hawk":

> . . . But that night I stole
> Off on the unbounded
> Beaches where the whole
> Of the Atlantic pounded . . .
>
> We have made a pass
> At the infinite,
> Made it, as it were,
> Rationally ours . . .

What better place than a beach for it all to have started? This first
symmetry between the shore and a runway, the ocean and the
sky, still holds. When a jet lowers its wheels over the water, when

it descends to solid ground from over the open water, it comes to land in every sense.

We have forgotten that the *Good Ship Lollipop* was an airplane. But when we transfer from the airport's supply of electricity to the aircraft's, an older captain may say we have switched *to ship's power*. In navigation terms we speak of *ship-derived* or *own-ship* positions. The captain is still the *skipper*, often abbreviated to *Skip* as a term of direct address—"Hey, Skip." As a copilot I am a *first officer* on an air*liner*, among the cabin crew are *pursers*. We talk of *forward* and *aft*; *cabins, galleys, bulkheads, holds, yokes; manifests, tacking, coamings*, and *trim*. We count aircraft by *hulls*. A colleague not sure if I am still flying the Airbus A320 or if I have switched to the Boeing 747, will ask me which *fleet* I am on. The small handle we use to turn the plane at low speeds on the ground, a sort of steering wheel that few visitors to the cockpit notice, is a *tiller*. Airplanes have *rudders*—and, in a linguistic twist analogous to those marine mammals that have re-evolved limbs better suited for their return to water, floatplanes may have *water rudders*.

The protrusions from the aircraft that hold antennas or drains are *masts*. The probes that measure the plane's indicated airspeed are called *Pitot tubes*, invented in the eighteenth century by a hydraulic engineer who studied Roman aqueducts and measured the speed of the Seine; who surely could never have conceived of the life his creation would come to in the latter-day skies above Paris, above everywhere. The naval heritage of the typical pilot's uniform— *pilot*, one who steers a ship—was chosen by Juan Trippe, the naval aviator who founded Pan Am and named his flying boats *Clippers*. The ambition of the 747's lead designer was to give his airplane the "stately majesty" of great liners; to conquer "oceans in a single flight." In the early days of air-traffic control, planes were tracked on maps using little weights cast in the shape of boats. Under the rules of the air, powered aircraft must give way to gliders, as on the water "steam shall give way to sail."

The flight computers can plot *abeam points* to tell us, for example, when we'll pass closest to a place—the only nearby major airport along a remote stretch of our route, for example—that we won't overfly directly. "We're abeam Luanda." Until now the Angolan capital's airport has only gotten closer, but from the abeam point it recedes. I sometimes catch myself using *abeam* by accident when giving driving directions: "You'll see the driveway when you're abeam the red silo."

I like the words for the patterned striations of cloud known as *herringbone* or *mackerel,* the ichthyology of our sea-sky. We speak, too, of *port* and *starboard.* There's a story that the term *posh* derives from a preference for cabins on the shady side of ships sailing from Britain to India—"port out, starboard home." It's false, apparently, only a pleasing tale. But this precise phrase is echoed in certain cockpits. Many systems on an airliner are duplicated, and we can turn a switch left or right in order to select which system the aircraft uses. It makes sense to use both regularly, if only to know when one of the two has stopped working. When I started flying the 747, the standard was to use the left-hand system when leaving London and the right-hand when returning to it: "port out, starboard home," the words of the legend printed right in the sober font of the flying manual. Starboard home, at least, remains a fine rule when sailing to Heathrow. If the wind is from the west the best prospects of London come to those on the right side of the plane, where copilots always sit, consoled by the view of what Churchill called "this mighty imperial city" as they wait to become captains.

When one pilot leaves their seat, they occasionally say to the other, only half-jokingly: "You have the conn": you have the ship, you are in control now on the flight *deck*. The lighter terms for the cockpit itself, when we answer calls from the cabin on the interphone system, all have a maritime bent: "This is Mark on the bridge" or "You've reached Nigel in the engine room." A friend who flies a small plane tells me he will not fly today because it is too windy over the hills near where he lives; he says the sky will be turbulent, the wind "like water over rocks in a stream." We refer to the turbulence that drifts off mountain ranges as *mountain waves*. North–south flights often cross the Intertropical Convergence Zone, a region near the equator. The trade winds converge here and the rising moist air fuels storms we must always watch out for, in the place known to sailors on the sea below us as the Doldrums.

Sadly, to me, we don't use *fathoms* for altitude in aviation. But our speed is measured in *knots;* what remains of us after our passage is our *wake*. When we arrive on a plane, we check its voluminous technical *log;* a jet without many entries is a *clean ship*. Such words remind pilots that the business of guiding vessels between blue-parted cities is an old one, and that our world is dominated by water, hardly less than that of the vessels that gave us language.

It's September 2002, and I am in Kidlington, north of Oxford. I've left my consulting company to start my flight training. In fact, I've nearly finished this training. I've completed all my written and airborne exams for my commercial pilot's license, but not without a last-minute hiccup.

Flight training is divided into visual flying, in which we look out the window to see where we are going; following mountain ridges and roads and railways along a map or through our memory; and instrument flying, in which we can fly safely in cloud, guided by the onboard instruments. Like many pilots who train in Europe, I was sent to complete my visual flight training in the generally cloud-less skies of Arizona, where the weather is perfect for it, and there are enough ranches for British instructors to use the same *Father Ted*–inspired instruction techniques they might deploy at home: "Push the control column forward, cows get bigger. Pull back, cows get smaller." After my visual training I returned to Europe, where the skies' tendency toward gray, cow-obscuring weather is equally well suited to instrument flying.

The most basic maneuver of this phase of training is an instru-ment approach. We descend toward the runway, typically following a radio aid on the ground. Then, just before landing, we change to visual flying. We land by looking out of the window, as we would if there had been no clouds at all—but only, of course, if we can see enough of the runway or its lighting to do so. At a certain mini-mum altitude, perhaps a few hundred feet above the ground, we must decide if we can see enough ahead of us to make this switch from the instruments to visual flight. If we cannot—if we are still in cloud, or snow or heavy rain or fog or whatever incarnation of high water might be impeding our view of the runway—then we abort the landing. We *go around*, we conduct a *missed approach*. Then we try again, or enter a holding pattern to wait for the weather to improve, or we fly away to land somewhere else.

For training, even in northwestern Europe, we cannot rely on thick clouds that terminate exactly at that minimum altitude. In order to simulate such weather *on minimums*, instructors must there-fore go to some trouble to blind their students to the outside world. In one of those oddly basic solutions that dot the high-technology field of aviation, instructors temporarily install a set of translucent screens all around the trainee's side of the cockpit, or require us to wear such *view-limiting devices* as a visor-like hood or *foggles*. When the *screens are up*, or we are *under the hood*, we can see only the instrument panel; nothing of the outside world. At the specified height the instructor will either remove the forward screen, allowing us to see enough to land, or they will leave it in place, forcing us to go around. On a twin-engine plane, at the point at which we abort the landing and climb away, the instructor may simultaneously reduce thrust on one of the engines to idle, simulating an engine failure at one of the most challenging points of flight.

During my final instrument exam, as we approached the minimum altitude for a runway near Bristol, the examiner reached up and began to move one of the screens. So I continued the descent for a second or two, as I could now just see the runway ahead and I thought he was in the process of removing the screens entirely. But he didn't. He then turned to me and said: "You have committed the cardinal sin of instrument flying. You have continued an approach below the minimum altitude without the appropriate visual reference." Crushed, I turned the plane back to our home airport.

The next day I repeat that portion of the exam with more success. I am at last in the clear, I think, a phrase that gets an understandable amount of airtime among pilots undergoing instrument training. But the following day my instructor calls. He tells me that there's a problem with my license. Though I've completed all my exams, I don't have quite enough hours in my logbook yet. We have to go flying, he says, "we have to go up"—for at least three hours and thirteen minutes.

"Where are we going?" I ask. "Wherever you like," he replies with a smile. It's an answer that aspiring commercial pilots, on a tight and expensive training schedule, do not hear all that often. So even before we took off it had the makings of a great day: a plane, a warm late-summer afternoon, blessedly clear skies all across southern England, no fixed itinerary. I was even allowed to invite a friend along.

We lift into the blue and head south toward the coast, following the edge of the Channel toward Eastbourne, Hastings, Dover. "Have you been to Canterbury?" the instructor asks. I have not. We bank to the northwest, near enough to see the cathedral. The very first cumulus clouds begin to form in the afternoon heat. We turn north toward the estuary of the Thames, crossing its waters with the sea on our right, while on our left the river, that in later years will become by far the one I know best from the air, winds upstream and disappears into the haze of the capital. We head across Essex and Suffolk, toward Norfolk and the Fens that might be the Netherlands, just across the sea.

Several American fighters from a nearby military base parallel us. It is like a Porsche racing a bicycle. Their extraordinary speed, as they rocket past us, shows us more than how slow we are; it gives the illusion that we are flying in reverse.

As we turn back toward Oxford we see the puffy billows of water are now blossoming all over southeast England, called up by the afternoon sun, as densely and randomly as dandelions in a field. For perhaps the most joyful half hour I will ever spend in the sky, I bank the plane left and right, frolicking and dodging through gaps that so many unseen calculations of the air and sun and earth have left.

I try to pin down what this sense of weaving reminds me of. The visual effect resembles a lower-stakes crossing of an asteroid field in a science-fiction movie. But such physical interactions with the clouds feel more like some aerial emancipation of downhill skiing, the quick alternation of dramatic lefts and rights, the sense of slipping forward even as I make sharp turns around a fleecier species of mogul. I turn back to my friend. She smiles and gives me two thumbs-up. The instructor, too, is apparently enjoying himself. After we land he remarks that such flights, with freshly minted pilots who are not under the shadow of formal instruction or exams, are a rare treat for him, too.

Since that day I've often been startled by the perfect joyfulness of flying close to clouds. The smallest clouds may be no more than tens of yards across. That is larger than the small plane I was flying, but smaller than a 747. Perhaps, for brains designed to maneuver us quickly through tight and dangerous places, it provides a kind of mischievous transcendence to dive between such white-hard arrangements of ethereality, near-absences that bear the visual weight of mountains.

It's an even greater pleasure to sail right through them. To fly directly into the billowing illusions of towering substance as if they were nothing—or as if it were us that were without corporality—is, to a new pilot, an entirely separate order of aerial pleasure. We close on structures that may be smaller than the plane, or the size of cities, clouds like sky lakes, their edges rotated into three dimensions of white shore; then we are amid the total, white nothing. That is, almost nothing. There is often just enough of a jolt when we enter a cloud, a quick rumble as we dive into the difference of sky, to remind us of the difference in sky-circumstance that explains why a cloud is there at all. Then we fly out the other side, and it all returns in the cleanest instant. There it is; the world, again.

Fluffy cumulus are the clouds I love best; they are flying high, on cloud nine, in seventh heaven; they are walking on air. They are the clouds that fill rococo artworks or that are painted upon faux skylights in the ceilings of the New York Public Library and Versailles. Even these, the most exuberant of clouds, seem to possess a slow, dignified consciousness, a feeling that's enhanced by our ability, from the windows of airplanes, to directly observe not only their movements but also their barely perceptible growth. If you have ever been on a whale watch, moving among clouds has something of the same quality—the sense that these enormous, lounging creatures are directed by steadier lights; that they can hardly notice something as small and jittery as us, and inhabit frames of time that we have lost.

Of course, 747s do not linger in playful summer cumulus that provides such delight to new pilots. Except for brief periods around takeoff and landing we mostly see these happy clouds from above, where the pleasure they give, like a memorable tale or joke, often takes the form of an unexpected reversal. Seen from cruising altitude, cumulus clouds—the connected parentheses or arcs cut from circles as children draw them—lie surprisingly close on the world. Like fireworks as we understand them from an airplane, they are basically ground events.

As we look down upon such clouds, upon what more typically looks down upon us, we realize that today it's the planet, not the sky, that is partly cloudy. This, I think, may be what I'll miss most after I retire—this customary, daily view of the gathered thoughts of our home skies; the low world beneath its eaves of water.

Often clouds form over land in the rising heat of the day, but fail to do so over nearby open water. On one memorable afternoon during my first summer on the 747, I flew west from London to New York. All of Cornwall, Devon, Somerset, and the southern Welsh coast were dotted with clouds, while the sea off these places itself remained as blue-clear as the sky above. In this way the clouds were a self-portrait of the land drawn by its own rising warmth, a mirroring cartography of mist, and also a kind of inverted sky-hourglass that measured out the lateness of the afternoon. Such cloudscapes often form over an island-dotted sea, where they form an aerial archipelago all their own, a map we can read long before we overfly the land it encodes.

Other days, though far from land, occasional cumulus clouds scatter above the open sea, each casting a small pool of darkness onto the oceanic blue. Here there's no earth to make sense of the patterns, no codex to the chaos of air instabilities that would answer the same question—why here?—raised by a lone tree standing in a meadow.

Contrails are a contraction of *condensation trails;* they can cover 5 percent of the busiest skies and may be described as man-made clouds. We see, too, their opposite. Sometimes a plane below us passes through the top edge of a cloud and the swirling heat of its engines will cut a trench right through it, and so the airplane leaves not a white trail in clear sky but a clear trail in a white one; a rare sight we might call an anti-contrail.

Sometimes an uneven wind, which we can feel in the cockpit is choppy, scatters the white contrails left by the airplane ahead of us, disassembling their rectitude and writing the turbulence on the blue in a freely tumbling white cursive. At other times a steady high breeze takes the contrail, immediately lifting this record of a jet's passage whole in one direction across the sky, away from the route we share; the drifting contrail is then a time stamp of the aircraft that made it. From the ground you can observe the formation of such wind-borne diasporas of contrails, when a plane passes and its after-cloud floats surprisingly quickly across anything high and steady in your line of sight—electrical wires, say, or the branches of a tree.

When there is no wind, the contrails stay in their appointed positions for some time, and so on a busy air route we occasionally see a stack of contrails, as neatly arranged as the lines of a pole fence. On a moonlit night, if there is only one above and one below, running off to the horizon to where each ghostly fog-line is headed by the glow of a distant plane, then it's as if the lights and the path of one craft are only the reflection of the other's.

I'm often disappointed when I arrive over a region of grand scenery—the American Southwest, Greenland, Iran—or any place I have not seen before, and find the world cloaked, or when the world below is cloudy for an entire flight. Such days, however, are a reminder that overcast days are not sealed off but divided. Regular flyers and even pilots may forget that to cross this division effortlessly, to sail into the upper hemisphere of our hours, is a new realm of experience. From the grayest of mornings, from the dullest of meetings and the longest of lines at the post office, we climb into the light-filled clerestory of the world.

If the plane is heavy and climbing slowly, or the cloud deck is ill defined, then the jet rises from the white as if lifted by some slow, innate buoyancy. The transition to the upper regions of the day is gradual. If, on the other hand, the plane is climbing quickly, and the tops are sharply defined, then the plane launches itself into the sky like a kickboard that a swimmer has forced underwater;

when it is let go and breaks the surface. Clouds are associated with disturbances and risings in the air; and so rising out of the clouds often means soaring out of turbulence, too, into a sky that is both clear and smooth; a sky that is clear because it is smooth.

There's a direct analogy between the career of a pilot and every flight that takes place on an overcast day. When you first learn to fly, you avoid the clouds. Later, you learn to use instruments to fly through them, and later still you may fly airliners that soar above them almost always. I remember clearly, then, the start of my instrument flight training. It was the first day I was permitted to fly through clouds, instead of air-skiing around them or remaining on the ground. It was also the first day I shared a radio frequency with actual airliners arriving in the skies of southeast England from all around the world; the first time I called "London Control."

It's surprising how many layers of cloud can lie over each other and the world, each with its own hues and personality of light; we climb from one to another of them as if rising through the storys, each so marvelously different, of a building cast of mist. Higher layers may be thin, even transparent, and we can often look through one layer of cloud at another beneath it. A high, thin layer of racing cloud has the organic and mathematical look of sand drifting across a hard-packed beach, or snow caught in the headlights of a car, blowing low over the dark road surface. At different heights such sheets appear to move at different speeds—world-scaled panels of water, sliding across each other, each as big as the sky.

The lowest of these surfaces may be the ocean itself, cut perhaps with the sharp angles of icebergs floating beneath a stack of mist-rounded insubstantialities. Like so many wonders from the window seat, such a sight feels both abstracted and true; there's no difference between how we might imagine such a scene and what we actually see.

Imagine the loveliest red sunset, and place it above the grayest of days, where we forget it so often is. Sometimes at dusk, climbing up through such layers of cloud, the plane emerges from the monochromatic low world into a dizzyingly smooth vault of nearly horizontal red light. As we rise in the sky it's as if the global circuit breaker for color had been tripped but has now been reset, as if this red is its own kind of cloud, yet another state of water, that we discovered in the heavens. The crimson cloud surfaces can then take on the look of some interstitial volume of the body, the inner tissues of a world without scale.

Georgia O'Keeffe was afraid of flying but obsessed with the clouds she saw from airplanes, which she painted with an all but religious devotion: "When you fly under even normal circumstances, you see such marvelous things, such incredible colors that you actually begin to believe in your dreams." I try to remember, when I haven't flown for some time, and the handles of the bags of groceries that I'm carrying through a cold and rainy November dusk are about to break, that such a lake of light may be above the clouds that rest upon the street.

In the normal visual pace of descent and arrival, possibilities of place narrow as specificities multiply. What we might call the arrival effect happens in two senses: the vertical and the horizontal. Arriving over land, the world resolves itself vertically, because more of its detail is visible as we descend closer to it. At the same time, the space between cities is transforming horizontally. Wilderness distills to farms, farms into suburbs that are riven by main roads that lead to the city itself. These twinned accelerations, these

parallel condensations of detail, are what it means to come to a
city from the air.

Place lag occurs because our sense of where we are cannot
travel as fast as the plane. The sight of an approaching city on a
crystal-clear day can briefly mask place lag, because what is gradually happening in the window has its own visual flow, a progression of geographic logic that may give the illusion of sensibility
or comprehension. But when place appears suddenly, only when
the clouds at last part, then for once what happens to place before
our eyes is aligned with what happens to it in our mind. The eye
closed, and now it opens.

On that flight to an overcast Amsterdam, the first flight I took
alone, I was disappointed to see so little from the window seat
during the arrival. But the Holland that emerged so late in the
window—the sea, the ships, the wind-lashed walkers on the beach,
the wet lanes of the morning rush-hour traffic bending toward
Amsterdam, the green fields, and the roof panels of the many
greenhouses—held my imagination for a long time after the flight;
much longer, perhaps, than it would have if the skies had been
clear. I came to love this gift from the clouds: the experience of
seeing so little for so long, until we see everything.

Some tropical cities, like Singapore, have been surrounded
by great vertical extents of rising afternoon cloud every time I
have flown to them. We descend into this world, we fly down and
around these columns of vapor long before we ever see the city's
skyscrapers or runways, as if we must pass first through the gates
and along the avenues of the cloud-city of Singapore before we
reach the concrete metropolis that is carrying on beneath it.

In London, the proximity and typical weather of Heathrow

mean that after hours of white and gray, often the first sight of
the returning earth is the heart of the city that called us across the
world, its wharves and new skyscrapers rising like great masts from
the ordinary busy morning that waited beneath the clouds.

We take such navigation for granted, as if it's nothing to us,
to fly across the planet and then to approach the white-granite
surface of the cloud world with the now ordinary intention of simply descending through it, to find all of London lying like pages
of densely typeset newsprint spread upon a floor. The sky waters
extinguished geography, perhaps for nearly the entire journey.
Now they part and place arrives, not as the long-building and
inevitable visual conclusion to a journey, but as if with the blinding, voltaic thump of light arrays high around a stadium: here is
London, in the toll of its present hour.

Low-lying fog, seen from the ground beneath it, all but extin-
guishes the world. From above, however, fog can resemble little
more than a veil of gauze drawn across the land, so low upon it,
seemingly, that for someone on the ground to rise above it would
be as simple as standing up.

Visibility measured along the runway is called *runway visual
range*. You can see the *transmissometers* that measure this near runways; they look like two periscopes that have emerged from the
earth, swiveled, and spotted each other and are now staring each
other down. A landscape's predilections for fog are an important
consideration when choosing where to build an airport, and so
transmissometers may be erected long before the airport itself. A
captain once told me a rumor about the residents near such a
potential new airport, who put garbage bags over the transmis-

someters, to conjure the illusion of dense fog at what, to distant engineers, must have suddenly seemed like the worst place in the world to site an airport.

Often when fog is causing delays that ripple throughout a day and across a continent, the sky above it is nevertheless perfectly clear. We prepare for a landing in fog while we are descending in clear and open sunlight. Only in the last seconds does the plane descend into the rolling waves of mist; the world and runway vanish as wholly as if someone has cast a gray sheet over the nose of the aircraft.

Sometimes the fog lies on a runway only in patches, and so we must evaluate multiple visibility reports from several different points along the runway before we decide to make an approach. Once I flew to Edinburgh and we landed in perfect sunshine. Then, a third of the way down the runway, we rolled into a total whiteout before, a thousand feet further, sailing back into bewildering billows of sunlight, an experience I otherwise associate with biking or driving across the Golden Gate Bridge.

Not long ago I approached London on a foggy autumn morning. A brighter lighting system had recently been installed on the southern runway we would shortly land on. The controllers directed us right over Heathrow, then east toward central London, before turning us back for our final approach. As I flew above the airport and banked overhead, I could see that the fog ran only in portions over this runway, in slow, wind-rolled breakers. It could have been a road through time, from a prehistoric moor to some future space age: one half of the runway was submerged in fog that blurred but did not entirely mask the lights beneath, while the other half was completely clear, its mathematical, illuminated patterns forming a dazzling welcome to the homeward-bound sky vessels.

Pre-dawn fog over the lights of a city itself is one of the most sublime things I have seen from an airplane. On such a morning the mist floats over the sprawling light-scape, piling thicker in some areas and thinning in others. The mist has the life of currents on a body of water or glaciers seen at high speed in their seasonal pulses over the land, water in slow motion, at perhaps its most archetypal and mysterious. Thinner fog can act on our vision like an out-of-focus lens, erasing the edges of even sharply demarcated light-structures. As roads blur into contrails of light and collections of houses spread like night flowers into the covering mist, the illuminations of an entire city can take on the quality of a string of Christmas bulbs on a bush covered in fresh snow.

When I took that examination flight at the end of my instrument training, I descended to an altitude where I had to decide whether I could see enough to land or whether to fly away from the runway. But the best-equipped airliners, flying to the best-equipped airports, offer another option on foggy days: an automatic landing. This is an extraordinary thing to see, because we see so little.

As we sail through the gray on an approach for an automatic landing, it is as if the volume of the world has been turned down. Partly it's our concentration in the cockpit and partly it's the peacefulness that many of us may associate with misty starts to autumn days. Such a sense of quiet is more than an impression— landing aircraft are more widely spaced in fog, so there may be fewer pilots speaking on each radio frequency. Foggy skies, too, are often nearly still, with little or none of the aerial terrain of bumps and turbulence that in higher clouds continually remind us that we are in the air. Often in a cockpit surrounded by fog there is no sense of motion at all, except in the turning of the altimeters. The feeling that fog can give, that the cockpit windows have been papered over in perfectly trimmed gray sheets, is one of the situations, unsurprisingly, for which a flight simulator prepares pilots most perfectly. Indeed, we may do more landings in dense fog in the simulator than in the actual aircraft, and so on the real world's mistiest days it's easy to forget that there isn't an examiner sitting behind us with a notepad.

The aura of silence on fog-bound approaches leaves room for an aspect of our return to earth that is not apparent to passengers: the airplane's own voices. The craft's most prominent voice is that which announces the final heights as registered by the eyes of the radio altimeter, the device that bounces radio waves off whatever is directly below the plane. These *callouts* annunciate our growing proximity to the earth. There are no altitude callouts on departure, when we are going away from the ground.

When I flew the Airbus the first of these announcements heard on an approach, in a low male voice, was "TWO THOUSAND FIVE HUNDRED." At the analogous point of flight in the 747, a female voice speaks the name of the device itself that has awoken to the radio-sight of the ground below: "RADIO ALTIME-TER!" After the wake-up call the radio altimeter begins its formal countdown with something of the majesty that I associate with the announcers at a rocket launch. The increments grow smaller, and so the frequency of the callouts increases as we descend, a quickening pace that perfectly reflects the ever-closer planet. Next is "ONE THOUSAND," followed by "FIVE HUNDRED." "FIFTY." "THIRTY." "TWENTY." . . . "TEN." A moment later, touchdown.

In the midst of this countdown comes another important voice. That minimum altitude to which we can descend without sight of the runway or its lights is called the *decision altitude* or *decision height*. This vertical milestone is so important that as the plane nears it, it will announce "FIFTY ABOVE" in the cockpit. We are not 50 feet above the ground; we are 50 feet above our decision in the air. The next call is known as the *decide call*. "DE-CIDE," says the 747 brightly but firmly. Can you see enough to land? "DECIDE" right now, choose between the earth and sky.

I first heard the decide call long before I became a pilot, when as a passenger I sat in the cockpit of an airliner for landing. At the time I was not long out of academia, not long into the business world, and it occurred to me afterward that many more situations in life—university seminar rooms and corporate meeting rooms, for example—should have such pre-programmed callouts. I now often mutter: "Decide," in the 747's unique accent and intonation, when I'm irritating myself by postponing small decisions in daily life. Even non-pilot friends to whom I have described the 747's decision-making strategies, when they have heard enough from me about some small dilemma, will say to me: "Decide!" The voice of the 747 that they associate with me is the voice I associate with a moment just before every landing, with hopes for the sight of approach lights running forward in the mist and murk.

After an *autoland* we must remember to disconnect the autopilot. If we try to leave the runway without doing so, the plane fights back; it tries to turn back through the controls, to keep us on the runway centerline. It doesn't know that it's we who have disturbed its trajectory; it only knows that it must stay where it was last commanded to be, in the center of the place it found for us when we ourselves could not see.

Taxiing in dense fog is much harder than flying in it. In noto-riously foggy places, such as Delhi in winter, a little car—a *follow-me* car—may be sent out to meet us as we leave the runway. The instruction from the controllers will then be to "fol-low the follow-me." The follow-me drivers are typically sportier than the pilots they are leading and so the car will often vanish into the fog ahead. We may not be able to see even the signs on the taxiways, let alone another plane, and so we bring the jet to a gentle but complete stop. The follow-me driver soon realizes that the 747 is no longer following and turns around to find us. Soon we see the headlights of the car returning to us in the mist, and we ask the driver from now on to please drive more slowly.

Water from above is fractal, abstract, almost without scale. It is impossible to sense the size of its features, or the inner life of those peculiar and marvelous surface flows of a different reflectivity and texture, which are the shimmering sums of local winds and currents, of tides and ship wakes. Water from altitude is metallic; a rare alloy of what the poet Mary Oliver called the "silver of water," from which sunstruck wings, too, so often appear to have been cut.

From above the open ocean we see the rippling pages of scal-loped blue, the waves that reflect on nothing, that will not break until their steadily spreading crests reach a far shore, crossing a volume as naturally as music from a stage or light from a star. Most waves on the ocean below are first conjured by wind, before becoming a new vertical dimension that the wind catches, an air-handle, that allows it to create yet greater waves. Indeed, a scientist tells me that the wind over the sea pulls up waves to greater heights as simply as the slipstream lifts the upper surface of the wings of an airliner. It's a pleasing thought, perhaps, that waves and wings are lifted by similar mechanisms of air; while the clouds that run in telltale patterns across the sky, in striped geometries of mist, are a reminder that there are waves in the atmosphere, too.

If you've ever been hiking or driving near the coast, perhaps there was a moment when you emerged from a dense forest or a curve in the road. The world suddenly opens, the floor of the sky falls away, and the paired blue vaults of the sky and sea stretch before you to their meeting place on the distant horizon. The chance to be present simultaneously below and above these wind-swept blues is, I think, what draws so many travelers to the world's sunniest and steepest coastlines.

Pilots based in Britain will cross the Channel regularly, each overflight a chance to think of Hubert Latham, the French aviator who first attempted to fly over the Channel, and as consequence became the first to land an airplane on the sea, where he lit a cigarette while awaiting rescue, and who died in Chad—killed by either a buffalo or a murderer, depending on which account of his death you believe. Early one bright summer morning, descending toward London from continental Europe, I saw an aircraft carrier and its battle group in the Channel, steaming southwest toward the Atlantic Ocean. The sun on the water crossed the long white lines of wakes trailing behind each vessel; the bar code of a navy, a nation, each vessel like a motorized word—*power, dawn, fleet.* We pointed the sight out to the passengers, several of whom stopped to talk with us about the ships after we landed. The term *blue-water navy* refers to a nation's long-distance, oceangoing vessels as opposed to a regional or coastal force. On mornings when I take off to start a long ocean crossing, I like to think: Here is a blue-water day.

When a plane departs from the airport of a port city and points immediately out to open water, it enters simultaneously the realms of the sky and of the sea. If our destination is also a coastal city, then we return to land in the same two senses. Similar, perhaps, to the effect of clouds on arrival, an air traveler's sense of return takes on a kind of purity when much of the descent takes place over water. Over land, the material of the view changes—fields, then roads, then factories. But in an approach over the sea, the content is unchanging—only its proximity continuously and frictionlessly alters. The eye follows these fluid transitions, over twenty or so descending minutes, from a blue abstraction to individual, three-dimensional waves, from the pleasing idea of water to the direct sight of its heaving surface.

My goddaughter once told me that she likes to have a globe because it reminds her that there is only one ocean. I think of this, of her, when I see more than one ocean in a flight. When I fly to Los Angeles from London, although we cross some portions of the Atlantic, its volume is diminished by thoughts tuned to a destination that stands on the edge of a further and greater ocean. Occasionally, at the end of such a flight we fly briefly out over the Pacific before turning back—only the first few miles of an ocean that, set against the mileage of the flight, is little more than an afterthought. Yet this ocean is the reason this city is here; it is the reason why we left London. After landing I may go to the beach, and there I have a vivid awareness that I am facing west, or thinking west, like the plane and the historical windsock of the city and America, that all the day's westward miles have brought us only to a beginning.

Standing on a west-facing beach, when the sun at last begins to set on the long day of a westbound flight, you may see a line of light that runs from the sun directly to you, dazzling over the water's surface. This effect is often seen from the cockpit or window seat, too, when the light turns and moves with you, connecting you to the horizon beneath the sun. It is sometimes called *sun glitter*. But to me it looks more like a road, a path or paving of light that connects your eye and vessel to the setting sun. It might be better named the *sun road*, or the *sun way*, or the *sun wake*.

It occurs with the moon, too—"moon wake," we might call it. I think it's most sublime not when it connects us to the moon over an intervening ocean, but when we are flying at night over a land scattered with many lakes, such as northern Canada. The moon wake doesn't appear at all when what lies between the plane and the moon is solid ground. But when a lake appears on the line between the airplane and the point of the horizon the moon is over, the moon wake flickers to yellow-white life over the interposed water. The moon wake ripples over the water until it crosses the lake's far shore and both light and lake vanish into the darkness, until the next of the lakes appears, each sounding off like notes strung along a stave.

In bygone days the edge of a continent was where travelers might rest or change from one sort of vehicle to another. Now it's the nature of long flights that the threshold of land may pass unnoticed. The boundaries of earth and water, the great divisions that have sculpted species and ecosystems, countries, and languages, are rendered inconsequential by an airplane.

Sometimes I mention to passengers when we will reach or cross which far coast, when we will "make landfall" over Ireland or Newfoundland. But this phrase—full of historical import, as if we were due to sight land at eye level, and to come ashore in small

boats, staggering through the surf clutching a text, lifting ourselves
onto the sill of a new world we had gone to an ocean of trouble to
reach—is never quite right in the air. In airplanes, even if the pilot
announces the moment, we cross over the rocky shores as if they
were fictions, boundaries of an expired empire, stone walls like the
old ones of New England that the forest has long since grown back
up around, the hairline fractures of unconsidered journeys.

It's early summer. I'm in my late twenties, not yet a pilot. I've been
working in the business world for a year or two now. I spent a
summer in Japan in high school, in Kanazawa, a charming city
on the Sea of Japan most famous for a castle and its nearby garden.
I lived with a Japanese family and studied Japanese at the local university. If you had asked me then if I hoped to live abroad someday,
and where, I would have answered yes, and Japan. (Later, when I
have the chance to move from a short-haul to a long-haul airliner,
flying to Japan is one of the dreams that make my decision so easy.)
    My boss in the consulting company has heard about my early
experiences in Japan. That is why I am assigned to this new project for a client in Japan, and why I am flying to Osaka today. To
get to Osaka from Boston I fly first to Dallas, the sort of check
mark–shaped itinerary that is consistent within the logic of airline
hubs, but that still strikes me as an arresting feature of our modern
journeys—that airplanes have so abstracted place that the great
logistical circle between New England and Japan should naturally
cross, say, the plains of north Texas. I stare out of the window
for the last half hour of this first flight of my itinerary, marveling
at the approach of a state larger than France, and then at the
sail-specked lakes of Dallas, dazzling in the sun wakes of the first
Texan morning of my life.

After a few hours working on my laptop in a café in the enormous airport, I'm embarrassingly excited to board an MD-11, a large three-engine airliner, for the first time. In flight I work again, eat lunch. I chat with the older man seated next to me, who tells me he has been doing business in Japan since the 1960s and will retire soon. I talk with the cabin crew about which restaurants in Osaka they'd recommend. One of the pilots comes out to greet some of the passengers. I ask him about the new airport in Osaka. I like the code for this new airport, KIX, I tell him. He laughs. It has been built on an artificial island in a bay, he says. A former naval aviator, he tells me it feels a bit like landing on a luxuriously dimensioned aircraft carrier. A few minutes later we each return to our work, to our computer screens. We are having very different days, I think.

Before I open my laptop again I turn to look out of the window. We are crossing a dizzyingly mountainous coastline, piled with forests and tawny, rocky peaks. It appears to be hardly inhabited. Where the steep land ends, where North America reaches both its horizontal and vertical conclusions, are long parenthetical surf lines, like the torn edges of expensive paper. This coast, the pilot later tells me, is California. Big Sur. We have a slightly southerly route today. We are crossing the entire Pacific.

It's surprisingly rare for an airplane to cross an ocean so simply and wholly, to say farewell to land at one point early in the journey and to have it appear again only near the end of that same journey. Even on a flight from the east coast to Europe, the definition of what we term a transatlantic journey, around half the flight may take place over land. Although we think of a flight from London to Seattle, too, as transatlantic, and culturally this is true, geographically and certainly visually it's better imagined as a journey of unweighted steps between the stones: Britain itself, the Hebrides, Iceland, Greenland. Baffin Island, the world's fifth-largest island, peaked with mountains named for Odin and Thor. The narrow, ice-clogged strait, the Fury and Hecla, that divides Baffin from the Canadian mainland and forms part of the North-west Passage. The Canadian Shield, the Rockies, the Cascades, at last Seattle. A grand journey; but hardly transoceanic.

Other island hops have a more inviting climate and scale. I once flew across the Aegean as a passenger, from Athens to Rhodes. There were so many islands that at nearly every point I could see water meeting a vertiginous coastline. Greece's archipelagos looked as if California had been hammered, shattered like glass and scattered across the blue, a blue so perfect that it appeared that not only water but the sky, too, was breaking along the new shores. It was easy to imagine how these coast-shards assembled to a history and a country, to a mythology, to the idea of a nation composed not so much of land as of its edges. On no other flight have I seen passengers pay such continuous attention to the view below.

Cape Town stands near the southwest tip of Africa. Only three-fifths of the northern hemisphere is water; but about four-fifths of the southern hemisphere is, a truth that alters the ring of the name Cape Town—or Kaapstad in Dutch and Afrikaans, iKapa in Xhosa. Cape, an accident of tectonics or erosion, a rocky hinge of history, a settlement at one end of its christeners' world, a reminder that the best place names may be those that make the most sense from above. The Cape of Good Hope, Cabo da Boa Esperança—though Bartolomeu Dias, when he first rounded it more than five centuries ago, twelve years before he was wrecked and drowned off it, called it the Cabo das Tormentas, the Cape of Storms.

It's the morning after a long night flight. I've had some cereal and coffee and now my colleagues and I are in the last stages of our approach to Cape Town. It's high spring in Europe but here it's autumn, a gray and blustery morning. The main runway at Cape Town's airport runs roughly north–south, not too far from the beaches of False Bay to the eastern side of the city, where no one at all will be swimming on a day like this. Today the wind blows gustily from the north, so the controllers send us past the airport and the city, past Africa, to the south. At some point over the water we will bank around completely, and after 6,000 south-bound miles we will come at last to Cape Town heading north, facing London.

The controllers instruct us to descend quite low, typically a clue that they plan our journey back to the airport to be short. But for some reason that we aren't told—perhaps another arriving flight has an ill passenger, or an animal has been spotted on the runway—we are instructed to maintain our southerly heading.

Between the clouds we catch the briefest sight of Cape Point, as we fly past the boxy terrain symbols on our screen that correspond to it, though of course nothing on the screen distinguishes them as among history's most significant rocks in the sky; to us they are not even named. Now we are over open water. Ahead are the foothills of the Southern Ocean; further still and unseen is the landless planetary belt of gale-tossed icy seas, the latitudes known as the Roaring Forties: Antarctica. The Cape, while not quite the most southerly point of Africa, is closer to the South Pole than Sydney or São Paulo. This unexpectedly long routing surely takes nearly all of the passengers to the southern zenith of their lives—a record that I myself would not break until I flew to Buenos Aires.

The flight is choppy in the squally winds. Rain showers pour from narrow, scattered towers of cloud. We fly through these streaming beams of rain, and then out into brief white-walled caverns of sunlight, through which brilliance falls and accumulates in small blue puddles on the gray sea. A moment later we are again bouncing through the rain and mist, then back into sun. We forget that such a scene, though I'd never seen quite its like before, is our home at its most ordinary. If you descended to 3,000 feet above a random latitude and longitude on the planet, this is what you would be most likely to find: battleship clouds over heaving ocean, no land in sight.

We pass near and size up a lonely freighter, tipping like a seesaw in the house-high waves, the thicker turbulence of its older realm. Finally we're instructed to turn: first east, then northeast. One last sharp turn as we lock onto the angled radio signal that drills up through the rain and mist, a beam we follow gratefully back to Africa.

Later, after a nap at the hotel, the rain has briefly stopped, and I have enough time before meeting the crew for dinner to drive down to Cape Point. It's a popular destination—height above water at its finest—and often I recognize passengers from my own flight there, though none on this leaden day. I walk up to the lighthouse. When I lean over the stone lookouts around it, I see seabirds screaming vertically up misty cliff faces, and I feel the wind as I have never felt it before. I stare down at the crashing waves and, beyond, the blue-and-slate kaleidoscope of sun and shadow, and the inbound tumbleweed of the next round of showers.

On a post, arrows point wildly around, each marking the distance and direction to a far-off place. Just-landed pilots, like recently arrived passengers, may look at such a tree of cities with a mild incredulity, and perhaps feel a particularly forceful gust of place lag. One arrow points to Antarctica; roughly opposite is the arrow for London. I think of timbered ships and the Cape's forgotten, stormy name, and I wonder if anyone was standing here by this lighthouse earlier today, to catch a glimpse of the 747 banking over the end of the Cape and the continent, or to hear the tuned thunder of the engines from within the wind-frayed clouds.

My father left the Congo in June 1958, flying first to Cairo, where he spent nine days, before traveling onward to Belgium. After further studies in Belgium he was sent to Brazil and for this journey he took a boat.

First, though, he took a train. As the sole passenger scheduled to board in Antwerp, he was asked by the shipping company if he wouldn't mind coming to Hamburg, saving their vessel a stop. The harbor and river of that city were locked in an icy January fog, he wrote in his notes, and throughout the first night onboard the foghorn sounded constantly. One night later in this long, diagonal journey down the Atlantic, rocky seas swept all the cutlery and crockery from the dining-room tables. Without a word the waiters cleared and reset everything, and then they poured great jugs of water out over the tablecloths, to hold the new service in place.

Flights, too, from Europe to Brazil often experience turbulence in the mid-Atlantic, where occasionally at night I see the light of a lone ship on the sea below, as if a star has misplaced itself below the horizon. I look inside the cockpit for the name of the 747 I'm flying, a city name typically, still sometimes written on a small plaque, and I remember the name of my father's ship, the *Santa Elena*, operated by the Hamburg Südamerikanische Dampfschifffahrts-Gesellschaft. A lifelong student of languages, he was pleased with a name like that; pleased that the company had not yet adopted spelling reforms that then favored the elimination of three con- secutive f's. He was already learning Portuguese, a language, he wrote, whose vowels seemed to enjoy pooling together as inti- mately as German consonants.

When my mother traveled from America to Paris, she sailed on the SS *France*—for some time the longest passenger ship in his- tory, built by a shipyard named the Chantiers de l'Atlantique. This name makes me wish we might someday christen an airplane fac- tory something like it. I work at the Seattle Sky Yards, an engineer might say, or at the Chantiers du Ciel in Toulouse. By the late 1950s more people traveled by airplane than ship between Europe and America. My mom returned to America in 1964, on an air- plane. It was the first time she flew.

The temptation is strong, but it's not quite right to think of airplanes as successors to ships. Passenger liners have all but dis- appeared and aviation, were it a country, would have the planet's nineteenth-largest economy. But there are more cargo ships and tankers than ever, steadily threading the world's cities under the aircraft that reflect the language and tradition of a maritime age that has not ended so much as fallen from the popular imagina- tion. Indeed, it is ships, even more than airplanes, that remind us that information technology and globalization are best regarded as separate revolutions which only occasionally overlap. There is nothing virtual about the exchange of physical goods, and far more of the world's ever-greater volume of trade travels by boat than by plane.

Airline pilots will see this more clearly than almost anyone. The crowded harbors we fly over are often splendid—timeless and sepia-tinged no matter how shiny the lines or computer screens of the 747, or how utilitarian the modern contours of the lower vessels. Flying near a port I sense exactly the historical continuity that is otherwise hard to find in my job; not a different world so much as a more classical version of my own. When I fly over the thriving boatscapes of Antwerp or Hong Kong or Long Beach, I remember I'm only one of the most recent to find work in the bonds between distant cities.

Boston, historically and still in its self-imagination, is first of all a port, and its harbor remains a busy place. In echoes of both Kitty Hawk and the city's maritime past, the busy runways of its airport stand so near to the water that mariners coming to Boston today might be forgiven for thinking they are about to dock at the airport; air passengers, meanwhile, might think they are on a seaplane, so late does proper New England ground appear in the window. Once I flew as a passenger from Shannon to Boston. We took off toward the sea and flew a southerly route that bypassed Maritime Canada before landing from the sea in Boston. Of the six or so hours of flight, the portions that took place over land amounted to less than thirty seconds.

When I lived in Boston I would walk to my consulting company's office on the waterfront in the North End, on the street named Atlantic Avenue. At the top of our building was a small glass-walled room where anyone could go, which we called the Crow's Nest. I would come here to work on presentations—or to spend a coffee break playing on the flight simulator on my laptop, before the wide view of the Inner Harbor and the airport beyond.

Years later, I'm a pilot, flying the 747 to Boston for the first time, within sight of my old office building. We sail south of the city and bank over the busy highways running toward the South Shore, then turn and line up for one of the northeast-running runways. As we descend toward the airport, the windscreen is crisscrossed by the many pleasure boats in the harbor, and a few sailing ships, their wakes twisting over the blue before the sober lines of the runways.

The city's venerable harbor, I learn today, has a formal presence in the cockpits of the most modern airliners sailing to the city. When we calculate the minimum height to which we can descend without sight of the runway, we must sometimes account for the dimensions of an unexpected species of aerial obstacle—the masts or structures of ships sailing near the airport.

Today the clouds aren't low, and we see the tallest sailing ship early enough, but even so we do not pass very high above it. When we are directly overhead, just when the broad lines of the 747 would most startle anyone on the decks below, we can no longer see the ship from the cockpit. Sails, too, are aerodynamic devices, wings of a fashion, and perhaps some gust of our passage descends to catch against their canvas; or our wake in the air echoes over the white trail their antique vessel leaves in the blue. Wood, metal, our shared dialect; from somewhere else in the city, someone has the right line across the harbor to see our wings cross above the billowing canvas, the bookends to the ages of Boston.

When I flew to Istanbul regularly, if the airport was busy, the air-traffic controllers would often send us on an extended tour of the Sea of Marmara. We could see clearly the situation of the ships in it, which were often stationary and had the appearance of waiting, as if for a Byzantine berth or an imperial audience. On a moonless night the water itself was without surface or depth, a matte-black mirror to the darkness above, and we saw only the lighted ships, a panel of scattered points steady on an unseen and tilted geometry, a night-bloom as magisterial as the eyes of waiting animals on a dark plain.

After landing we would go to our hotel, a dark, glassy sky-scraper on the seafront. From a high floor and through the smoky windowpanes, the waiting ship lights seemed to hang in the sky, forming a kind of vessel-lanterned gate to the Bosporus. The lights of flights later than my own, winding across the pane before turning back toward the airport on the European side of the waters, then appeared to move among the vessels.

In so many languages the word *airport* equates to *air harbor*—*luchthaven* in Dutch, for example—something we may not immediately hear through our overfamiliarity with the English word, our "ports of the air," or even in the pleasing tautology of Sky Harbor International airport in Phoenix—a name, though the landlocked city is surrounded by desert, that from the sky seems entirely at home.

You see many ships in the North Sea bound for the vast port at Rotterdam, as I did on that flight to Amsterdam, when the ships were the first thing I saw, crossing as if in the clouds. Nothing speaks quite so well of the Netherlands, of its character and more or less unending mercantile age, than the multitudes of ships that approach it at all hours, under the wings of your own ship descending to Schiphol—*Ship*-hol by some etymologies—the sky port where we land below sea level.

At the far end of Eurasia from Amsterdam stands Singapore, the fortress founded by Stamford Raffles, to challenge the far reaches of the Dutch Empire. (Raffles was born at sea. I do not know what paperwork was required off Jamaica in 1781, but we have a cockpit form to complete in the event of an in-flight birth, which asks for the time of birth in GMT and only the approximate position of the aircraft over the world.) Each time I land in Singapore I'm astounded all over again by the scale of maritime traffic in the waters around it. It's a common place for ships to pass by, as well as to dock. From the air we see clearly what Raffles understood, that this is perhaps the most obvious place on the planet to lay the first stones of a trading post, a port, a great city.

From above I do not doubt what I have read, that something like one-quarter of the world's trade, and an even larger share of the world's sea-transported oil, passes through the Strait of Malacca. These fabled, shallow waters have given their name to the largest class of vessel that can pass here: Malaccamax. Sometimes I fly to Singapore only a few days after a flight to a city on the Persian Gulf, above which it can be so hazy that it is sometimes hard to see the water at all, and the many tankers pointed in all directions appear to be flying, too, with the stately awkwardness of space-ships on a movie screen. Perhaps, the sight of a vessel approaching Singapore reminds me, I flew over this ship somewhere else; or perhaps the fuel in the tanks of this 747 moved in the holds of the very ship steaming steadily below it.

Close to the airport, in the Strait of Singapore, there are so many ships that you can begin to lose the awareness that you are looking at a waterborne scene. The effect is hard to grasp, as if someone had scattered hundreds of matchboxes over the kitchen floor. You must be mistaken, you tell yourself, something so numerous must be something smaller than ships. The planes above, their far-born, nearly-complete journeys winding over some of the planet's most crowded waters to the busy runways of Changi airport, mirror the stately chaos of commerce below. I do not know of any place where the history of a city and such an aerial snapshot of our age overlay each other as perfectly as they do here.

At Singapore, as at Boston, as at Copenhagen and Bermuda, our cockpit procedures are harmonized not just to the language of ships but to the height of the present-day vessels on the water below us. Our arrival charts here warn of high *maritime vessels,* a term that for nearly all of recorded history would have been redundant but that now draws a pleasingly necessary distinction for the pilots of airships. At takeoff for the return flight to London, the plane is inevitably heavy with cargo and fuel for the longest route I fly. But our takeoff power setting may be raised yet higher to account for the height of ships in the waters around Singapore—by their *air draft,* the path they make across our shared threshold to the sky.

Snow and airplanes may not be the best of friends, but they work together much more smoothly than snow and airports. The most formidable snow-induced challenges in aviation, in my experience, take place not in flight but on the ground, on taxiways after landing or before takeoff. Several times I've made a routine landing in moderate snowfall, but on the way to the terminal we've been forced to a complete stop for half an hour or more, prevented from proceeding because the taxiways are too icy or because we are unable to distinguish the tarmac from the grass, in the same way that from high over the Arctic we often cannot tell where frozen land ends and frozen sea begins. Airports are open, windswept places; it is difficult to keep runways and taxiways clear, especially in the strong winds that often rise as the snow stops falling.

The storm-tossed vessel in Coleridge's *Rime of the Ancient Mariner* is driven toward Antarctica, the "land of mist and snow," where "ice, mast-high, came floating by." Such imagery, of ice in the sea or ice forming on a vessel itself, conjures an extreme sense of wayfaring; we have traveled so far across water that the water itself is changing form. A pilot may think of Coleridge's title when learning about "rime," one of the many forms of ice that pilots will study. There is also *hoar frost, active frost* and the dreaded *clear ice;* there's *freezing drizzle, ice pellets,* and *freezing fog.* The definitions are both precise and ordinary. Snow *pellets* typically bounce and sometimes shatter. Snow *grains* do not. The definitions evoke playground training: *wet* snow forms a snowball, *dry* snow will fall apart if you try.

My visual flight training near Phoenix ran from late autumn through early winter. To make the most of the daylight we would start our preflight preparations before dawn. Often, even in that famously warm city, thick frost would build up on the wings during the cold desert nights. When the sun finally rose and hit one wing, the frost on it would melt in seconds, hardly less quickly than a hairdryer clears a misted mirror. Then we would untie the plane and push it, turn it around, so that the new day would fall on the other wing, and cleanse it, too, of ice. And then we would be ready to fly.

Occasionally we encounter ice in flight. On the Airbus I flew earlier in my career it was easy to see when ice had begun to form on the wings. But there was also a tiny pole—what we might call an ice catcher—perched outside the front windows of the cockpit, which served us like a canary in a mine. If we saw ice form on this pole then we could assume it had formed elsewhere on the aircraft. The probe contained a dim light so that we could examine it at night, but I found it easier to shine my flashlight forward through the window into the freezing slipstream, to see whether ice had gathered on the finger the plane held forth into the night for only this purpose. In the blackness, where despite our speed there was often no visual sense of motion, the plane felt like a deep-sea probe, with me peering out into the night from behind the thick panes, shining a light entirely out of scale with the enormous watery volume it barely penetrated.

The equipment to remove or prevent icing on wings surprised me when I learned to fly. It says something appealing about wings, and about speed and air, that this equipment is installed only around the front of the wings, their leading edges. In general terms *super-cooled water entrained in airflow* does not accumulate on the tops of the wings during flight; it does not even touch the tops, as if in deference to how perfectly the wings part the air. Only after landing, when the plane slows and the wings are no longer wings, will they begin to whiten beneath the falling snow.

Engines also have de-icing systems. Unless the air is unusually warm or cold it's assumed that any cloud—any *visible moisture*, a formal term that includes rain, fog, and snow as well as clouds—can cause icing. On the 747 this system typically works in an automatic mode in flight, but on the Airbus we activated it manually. We would turn the system on nearly every time we flew into a cloud, and off again when we flew out of it. The pressing of the buttons became as ordinary a ritual as turning on a car's windshield wipers when it starts to rain—an action to perform when the world turns white, and again when the world turns blue.

In weather forecasts in the American West you sometimes hear the term *snow line* or *snow level*, followed by an altitude; this is the horizontal division of the sky where snow turns to rain, a term and concept that makes particular sense from above. The snow level appears on the mountains, like the waterline against the depth marks of the side of a ship—a calendar, a slide rule that descends in winter and rises in spring. Often I land in a city in snow and walk through it that night, or the next bright morning, a city transformed, and it occurs to me that the snow and I descended together. At other times we fly through snow but land in rain in a place without mountains, crossing the snow level that weather forecasters would enumerate, if only there were hillsides here to reflect it. For those who like snow as much as me, it's a pleasure to imagine, when seeking shelter from a cold winter rain, that not too far above me a blizzard may be raging.

At night, mountains without snow are shadows on shadows. But snowcapped mountains glow even in starlight and in moonlight they come alive as vividly as cumulus clouds do—ghostly cones, divine blankets cast silently over unseen forms. There are mountainous lands such as Afghanistan and Pakistan that I have seen almost exclusively at night, their snowcapped highlands striped with zebra-like dark valleys where snow has not fallen or has already melted. Even flatlands show a pleasingly different face when covered in snow. When I think of Minnesota, for example, I think first of flying over it at night in winter: steady glowing cities luminous on the snow, under the moon and stars, a land and season that are never truly dark.

Heavy snow, especially at night, greatly impedes our ability to see ahead, so much so that, as in fog, controllers may issue visibility reports from the transmissometers on the runway, and automatic landings may be required. In the first part of the descent the airplane's strobe lights illuminate the composition of the storm, the way a flashbulb spotlights faces in a dark and crowded room. Each flash locks the snowflakes, lifted in the wind and racing past the plane at hundreds of miles an hour, into a seemingly impossible still image, a frozen moment in the inner life of a blizzard.

Later in the descent the steady, forward-facing landing lights may be turned on and so the flashing, time-freezing effect of the strobe lights is diminished. Unlike rain, which appears from nowhere on the surface of the windows, if we see it at all, snow-flakes in these beams appear as actual objects, as a new storm of spotlit ghostly flakes that fly continuously toward and over us. And so it is snow that gives us the rarest glimpse of the aircraft's true speed. Snowstorms, after all, are the only time visible objects are so close to us in flight. The racetrack pace of the streaming snow is like nothing so much as the graphics used to indicate fast travel in science-fiction movies—stars that motion turns to perfect white lines across the darkness.

Sometimes above Canada we see temporary ice roads, drawn over frozen bodies of water for vehicles and their brave drivers to cross. The ice roads often form straight lines, and the eye, sur-veying an hours-long chaos of wilderness, is instantly drawn to the unnatural sight of anything straight. They often mirror the contrails a jet draws above them in the sky, which like ice roads are straight and identifiably man-made, at least until the wind starts to work on them.

Once, in Helsinki in midwinter, a captain and I walked down to the quiet waterfront in a blisteringly cold wind, because a waiter had told us that even on such a frozen night the ferries were still running, and neither of us had ever been on an icebreaker. In the nearly subarctic darkness we boarded the ferry bound for Suomenlinna, the great island fortress in the city's ice-caked har-bor. The vessel was as quiet as the city. We told the captain we were pilots and without a smile or a word he motioned us to ride with him in the wheelhouse, from where we watched the nearly empty ferry bump through the jet-black water, casually knocking

car-sized chunks of ice that tumbled off to the left or right of our
course. Visually, the effect was similar to a flight among cumulus
clouds, but one jarred by the distinctly un-vaporous thumps of
solid ice that the bow cast aside. It was easier, the ferry captain
said, to follow the trail he had forced earlier than to make a new
one. The marbled path through the solid white was the inversion
of an ice road; it was a trail broken in water.

Later that winter on a clear day I passed to the south of Helsinki,
over the Gulf of Finland, en route to St. Petersburg. I saw from
high above the ferry-cut paths similar to those we had seen in
Helsinki's harbor, but on a much larger scale. The ice-crumbled
highways running in lazy arcs through the frozen sheet of the gulf
were left by large Baltic ferries. The lines formed a life-sized chart
of the ferry routes; they had exactly the shape and perfect sweep
that you might see on maps of early undersea telegraph-cable
routes or of the idealized paths between cities that appear in the
back pages of an airliner's on-board magazine.

The over-the-top truth of great circles means that planes routing
between otherwise balmy cities—Tokyo and Atlanta, Dubai and
Los Angeles, flights on which no one has brought winter gloves—
typically cross over the far north. The congenital chilliness of great
circles is apparent in the southern hemisphere, too, though far
fewer airliners ply them. Once in Buenos Aires, between flights
from and to São Paulo, I saw another 747, bound for Australia.
The captain and I, both far more accustomed to the geographies
and routings around the opposite, north side of the planet, made a
bet about whether the so-called straight line between Buenos Aires
and Sydney, such famously sultry metropolises, reaches Antarctica.
It nearly does.

WATER   215

As autumn turns to winter an enormous portion of North
America and North Asia, where many long-haul pilots spend
much of their workday, falls into whiteness. For pilots as well as
for passengers, these hours over this cold realm—on an entirely
routine flight between the worlds known as Los Angeles and Paris,
for example—are an opportunity few of us will otherwise have
to meditate on temperatures and places we will never stand in.
The room-temperature cabin of the plane arcs over lands and seas
masked entirely by the white that Melville described in *Moby-Dick*—
of the whale but also of polar bears, and ghosts, and horses of legend, and the "vast archangel wings" of the albatross—a whiteness
that was "the monumental white shroud" of Arctic lands, or the
"phantom of the whitened waters" that brings forth "a peculiar
apparition to the soul." A whiteness, in other words, as with that
of clouds, that is reason enough to reach for my sunglasses.

My father once remarked that growing up in Belgium, it was
possible to discern whether someone was from only a village or
two away by their accent. When I fly over a populated, temperate
part of the world, it is easy to look down at the kinds of vegetation, and the terrain, and to imagine that—at least before modern
nation-states and their education systems—languages once flowed
gradually from place to place, changing as naturally as ecosystems, words leading on to words. In aviation we speak of *isobars*,
lines of constant air pressure on a map; *isotachs*, lines of constant
wind speed; and *isogons*, a general term that we most often use for
lines of constant magnetic variation. An *isogloss* is the geographic
boundary of a characteristic of language—the borderline of the
natural range of a word, an accent, a feature of syntax.

Flying over the populated realms of Europe or Asia, I may look
down and ask what the language is here; how words and sounds

change as this land rolls into another. Sometimes the question is answered. We hear the accents of controllers change as we switch from London controllers to their Scottish or Irish colleagues; as we cross between Quebec and the rest of Canada; when we cross the U.S.-Canadian border, and then when we move across the United States, especially from north to south. But over the remote parts of the far north—places uninhabited, or inhabited so lightly that such inhabitation is as invisible on the land as in the modern imagination—this visually inspired question about the sound of a place does not arise, and the controllers we speak to from over such places may themselves be very far away.

Occasionally there are names to hear, the beacons of small places or geographic features that we can scarcely discern in the white. Once, over Siberia, I saw a river, its motion frozen whole upon the land. At home I looked it up; it was the Lena, from which Vladimir Ilyich Ulyanov is said to have taken the alias Lenin, as if Lincoln or Churchill had adopted some version of Mississippi or Thames. Spring leaves unexpected marks on Siberia. The southern portions of the rivers melt first, but many of them flow north, toward the ice dams, where the river has not yet thawed. Spring piles up; the liquid season floods the land.

Climate scientists, who have the best reasons to look down at the realms of cold water, may rely not on satellite photographs but on more specialized, satellite-imaging tools to fully distinguish clouds in the sky from ice in the sea. From airliners, the challenge they face is obvious. Often the sea off the Labrador coast of Canada will be filled with chunks of ice so numerous, and so small from the height of an airliner, so hauntingly gathered and conducted by what looks as if it must be an aerial force, that their pixels run together to form yet another kind of cloud. Only when

you look closely might you see that the bleached curves and contours of these surface nebulae are composed not of clouds but of
tiny imperfect discs of ice, swept along as if they were nothing
larger than flecks of dried house paint scratched from your hands
into the kitchen sink.

Sometimes you see a line of blue cut straight through such a
cloud of ice on the sea, and you follow it with your eye, certain
that so true a course must end in the shining steel of something
man-made—an icebreaker, surely? Yet the blue trail concludes not
with a ship but with a large iceberg. So much of an iceberg's volume is underwater that the winds that gather and guide the streaming puzzle pieces of sea ice may barely move the berg that parts this
surface flow; and so the iceberg slices or casts a kind of ice shadow
of open blue water behind it, a jarring exhibit of the superficial,
cut cleanly by depth.

I've heard many pilots say that their best-loved sight in all the
world is Greenland, which long-haul pilots overfly regularly on
routes from Europe to western North America. We reach this most
dramatic of coasts three or so hours into a flight between Europe
and western North America. The overcast skies of Scotland and
Iceland have usually cleared; indeed, the clouds often vanish just
as we approach Greenland's eastern, all-but-vertical shore.

On our screens, on our *terrain display*, the snowy mountains of
the Greenlandic coast grow in digital splendor not long before
they appear in the window, rising from the ocean like a skyline
approached from across a harbor.

The ocean waters that run up to these hundred coastal Swit-
zerlands may be sheet-white, frozen solid, or a liquid, neon blue.
Amid the sea ice, or alone in the open blue, are white constellations of newborn icebergs. I like to imagine what we are too high

to hear: the thunder of icebergs calving from the glacier's flowing edge; the rolling roar of the new icebergs' sudden overturning; the steady drip of melt from overhanging edges onto the sea, the unlikely percussion of the rain that falls more heavily in sunlight. Some icebergs are so vertically dramatic that even from a cruising airliner you can see their rising shape, and they are tall enough to shadow themselves; the fraction of them in the air alone is enough to remind us that *berg* means mountain. Hours later, sitting quietly at a desk by the open window of a warm and sunset-lit California hotel room, confounded by the memory of the ordinary lunch I ate over Greenland, I may look up the names attached to the world I cannot quite believe I overflew: *fast ice, second-year ice* and *tide cracks; nilas, ice keels* and *polynyas;* the settlements known as Ilulissat, Upernavik, Thule.

We will never see the water gyre of the planet as clearly as we do from above Greenland. Sometimes the skies are clear over the glaciers flowing through the coastal mountains, while low cloud lies over the open water where the glacier ends and we see only a continuity of white streaming down the fjord, which along some blurred line transitions from a river of ice to a river of cloud. Further in on the inland ice we see the sapphire eyes of melt pools that run out to rivers the color of the sky. If the skies are clear over the ocean, too, then we see icebergs beginning their journeys across the blue that will be their end, and no end at all.

Though Greenland has the form of a bowl of mountains, often we see little of the rock rim itself. Here at the edge of the bowl the Platonic angles of the cloud-swirled peaks are nearly pure folded snow, so much snow and so little rock that we perceive the land only as degrees of light, as a planetary drawing class on advanced techniques of shading; as crumpled pages of white earth smoldering in incendiary sunlight. And this is all, really, that the eye demands of a mountain: white shadowing white, the snow's idea of height, gracing the twinned voltaic blues of the sky, and the ice-clouded sea. We say we love the sight of this place, this land above all others; yet all we see is water.

Encounters

I'm in my midtwenties, on a business trip for my consulting company; there are several years to go yet before I'll become a pilot. The first plane ride I can remember was a family trip to Belgium, when I was seven. Now I'm further than ever, it seems, from that wide-eyed boy: I have a laptop and a stack of freshly printed business cards, in different languages on each side, and a garment bag filled with the suits I'll need on this long journey away from my office.

I can't decide whether to ask if I can visit the cockpit. From childhood through my college days, I regularly asked to do so. But since starting work, I've made such requests much less often. Partly it's because my colleagues and I often have to work on the plane, or we try our best to sleep before the meetings that are waiting for us in the morning after we land. Perhaps I also fear that my eagerness about airplanes might come off as unworldly or unprofessional.

Still, this business trip is special to me. It's part of a journey that I've pondered for weeks in advance, poring over an atlas in the apartment I share in Boston. I will think of this trip for years to come, whenever I see an image of the earth from space or encounter a photograph of my bedroom as a child that includes the globe

I owned then. The journey I'm on runs from Boston to Japan, where I will stay for several weeks, then on to Europe, and finally back to New England. I'm flying around the world.

The industry I work in, management consulting, is known for asking candidates for employment questions to which they probably will not know the answer, in order to see how they reason their way to a sensible guess. "How many trees are there in Canada?" is such a question, and one that in subsequent years I will have no shortage of time to ponder during flights over that country's boreal forest. In my own interview I was asked to estimate the number of violins in America, so I thought about how many violinists there were in my school, a figure I then tried to scale up to the country. Once, when I myself conducted such an interview, I asked a candidate to estimate the percentage of the world's population that had ever been on an airplane (roughly 80 percent of the U.S. and UK populations have flown at least once; worldwide, there are no statistics, but I suspect the portion of humanity that has flown is well under 20 percent—the percentage, incidentally, of Americans who had flown in 1965).

Another such question might be how many people, in all of human history, have traveled around the planet? This elemental motion, from home back around to home, returning without turning, remains rare among even the most seasoned air travelers—rare among even pilots.

Now I am flying the long middle leg of this journey, between Tokyo and London, on a 747. Before we boarded I could not hide my excitement. Even last night on a high floor of a hotel in the Shinjuku area of the city, when I looked out at the darkening sky over the lightscape of a city that is like no other, it was London as much as Tokyo that was on my mind; twelve hours in the air, 6,000 miles, from one island nation's great metropolis across nearly all of Asia and Europe to another's.

We've been flying for maybe five hours now. The world outside is entirely white. Land, not cloud, I think, but I can't be sure. We're somewhere over Siberia. I have never been over Siberia, and most of the other passengers are sleeping, their blinds closed against a day they won't even think about beginning to end until hours after we land in London. The next time one of the flight attendants passes me, I close my laptop and ask if it might be possible to visit the cockpit. She returns a few minutes later. Come with me, she says with a smile. I follow her upstairs. It's the first time I've ever been in the cockpit of a 747; it's the first time I've ever been upstairs on an airplane. I would not believe it if anyone told me that not so many years later I would fly this very plane between these same two cities.

The flight attendant introduces me to the pilots, who invite me to sit down. One of the copilots asks me about my work, but I'm much more interested in talking about his. He describes the challenges of long Siberian flights. He points out the magenta string of our route, arcing up to the top of a navigation screen. He shows me weather reports that print like receipts from the center console, enumerating the all but otherworldly temperatures of some of the Russian cities we are flying near. He talks with a combination of amazement, amusement, and acceptance about the peculiarities of the days and nights of a pilot's life—the oddly casual sense of going to Tokyo for the weekend; the challenges of managing rest before, during, and after such a journey; the vagaries of light, the twenty-four hours everyone on this plane will have between dawn and dusk, a dilation of our shared day over a fair portion of the earth's landmass. The captain shows me the print-out of his schedule, folded and stored inside his cap, a tradition I'll adopt myself years later. The codes and times on this sheet tell him that a week from now he'll be in Cape Town, then Sydney ten days after that. Twenty minutes or so later, aware that my enthusiasm might lead me to overstay my welcome, I reluctantly thank them and excuse myself.

I return to my seat, work a little more on a presentation, gaze out the window, doze. A few hours later another flight attendant comes to my seat. I've been invited, she says, to return to the cockpit for the landing at Heathrow. Would I like to go? I am out of my seat before she finishes the question.

I'm given a headset. As we speak a city rises in the windows above the computer screens, its miniaturized perfection turning steadily on the drum of the sea. I point to it. That's Copenhagen, on the Sound, the Øresund, the pilot says, smiling and making a slash in the air, between Denmark and Sweden. I try to remember the name of the place on the coast near the city where Isak Dinesen was born and died. Copenhagen is a city the copilot recognizes by sight; a city that means he is almost home to England. Here in the cockpit I first see the world as a place where the miles between Copenhagen and London are all but an afterthought, the endpapers to a day of work in the sky over Eurasia—a place where an entire city shines up, its name and situation on the planet read as easily as the sign for a highway exit after a long drive.

The captain points out the arc of the Frisian Islands, off the north coast of the Netherlands, and I remember one of my favorite books as a teenager, that contained brief histories and sample texts for hundreds of languages. The entry for Frisian, a language I had not heard of until I read this book, described how closely related it is to English. I hear a controller tell the pilots to "call now London." As if the plane is tracing the progression of language as closely as the waypoints that form the route—as if to see and hear this from above was the only purpose of airplanes and radios—the voice of an English controller soon begins to direct our descent.

I have never before sat in an airliner's cockpit for landing. Much of what I experience this afternoon will continue to amaze me for years to come, even after I am myself a pilot: the dramatic siren when the autopilot is disconnected—for the now-obvious reason that it should never disconnect without the pilot knowing—and another auditory marvel, those voices of the plane that crisply announce our heights as the runway approaches. At 200 feet, fifteen seconds above Britain: "DECIDE."

Even more striking than the autopilot and the voices are the earlier portions of the descent. I see, for the first time from the inside of a cockpit, something of the nature of airliners, something that I first began to understand from that Saudi plane I watched as it parked at JFK so many years ago. I see what a 747 has done to a hazy and half-remembered morning in Tokyo: suddenly we are among the scattered afternoon clouds that were below us; they are billowing past and over us until London is recalled from beneath them. I have loved flying for as long as I can remember, and yet until today I didn't even know what it is to be a pilot, that a job existed in which the sight of a city could grace a day so simply.

Four years later, I've become a pilot. I am walking in the airport in Los Angeles, about to fly to London as a passenger. Suddenly I spot the copilot who was so kind to me on that flight from Tokyo, who went out of his way to show me something I will remember for the rest of my life. I call out, say hello, and explain from where I think I remember him. After a moment of hesitation he recalls our previous meeting. We speak for some time. He congratulates me for having joined his profession and company in the years since we last met. Then, for the second time, he flies me to London.

Three years later I am in a bar outside Tokyo, or perhaps a café in Beijing or Singapore. I see him and say hello. I have now just started to fly the 747, which he still flies. In this sense my journey feels more complete than at our previous meeting. We talk for a few minutes, then say good-bye. When we next meet it will be at a *churrascaria* in São Paulo, several years later again. We share a meal together, an amiable chat, then bid each other farewell, until some other year and city.

If my connection with this colleague is so memorable, it's in part because it began with the first landing I ever watched from the cockpit of an airliner. But it's also because such an ongoing personal bond, although to outsiders it may sound like hardly a bond at all, is relatively usual. If aviation overturns our locally grown senses of time and place, it changes our sense of community as well. For many who work in this business of connecting people and places, the nature of our work means that many kinds of connections are not possible, while other friendships are valuable precisely because they are so rare, attenuated over both time and the fullness of the planet.

I'm often asked what actually happens when pilots go to the airport—where we park, if we drive to work; how long before a flight we're obliged to be at the airport; whether we meet our colleagues somewhere before a flight, or if we come together only on the airplane.

When I arrive at the airport for work, I first check my bag in, if I'm going on a long trip. Then I go to a large office area that stands on its own level between the arrival and departure floors that are so familiar to passengers. I sometimes take the stairs, a chance for movement before half a day spent all but motionless in the cockpit. Roughly equal shares of the floor space in this office area are devoted to computer terminals, meeting rooms, and a bustling café. At a computer terminal, coffee in hand, I scan the notices that have been published since I last came to work. These may relate to a new procedure or piece of equipment—for example, when a new computer system was installed on 747s, we were informed via such notices about the technical modifications that had been made to the aircraft and the changes required to certain cockpit procedures.

At some point, I must also remember to formally register that I am actually at the airport, via a swipe of my ID card. Otherwise at my *report*—the time I'm due at the airport, typically ninety minutes before a long-haul flight—someone will call my cell phone to check that I'm not stuck on the side of the road with a flat tire, or drinking coffee on the couch at home, having misread my schedule.

At report time, I head to another computer terminal outside a designated meeting room. Here I meet the other pilots and greet the flight attendants. Friends often assume that I work together with a fixed set of colleagues—a few pilots and a group of flight attendants, a more or less permanent team. The reality, at least in my corner of the profession, could not be more different. The total number of pilots and cabin crew on a 747 flight may be sixteen, occasionally as many as twenty. When I arrive at that meeting room, it's likely that I have never before met any of the people with whom I am about to cross the world. Our nametags are not worn only for passengers.

Our preflight briefing has two components. In the joint portion, we speak with the flight attendants about our journey together. They tell us what is different in the cabin today—perhaps a large group of blind passengers are onboard, or a royal family, or, as on a recent trip of mine, several hundred police officers traveling for a sponsored charity run.

The most relevant details which we give to the flight attendants are the time, and whether and when turbulence might be expected—both determined to some degree by the forecast strength and location of the high winds. We also speak about the route, and whether we will cross any remote areas of the world, such as Siberia or northern Canada or the mid-Atlantic Ocean, or mountain ranges that would change our procedures in the event of problems with the cabin air. We discuss any specific precautions for our destination, regarding malaria, for example. We check if anyone is bringing friends or family along with them—passengers who, because they accompany a crew member, may be affectionately referred to as *Klingons*. Not least, we check that we all agree on the location of our aircraft—no small matter in a large airport.

We may then discuss recent changes to our manuals or a specific safety-related scenario. The most fruitful discussions take place around situations that have different implications on opposite sides of the cockpit door. For example, pilots have elaborate procedures to follow in the cockpit in the event of a loss of cabin air pressure. The flight attendants, too, have procedures for this. The coordination of these two, at a busy time and across a large aircraft, speaking behind oxygen masks over mountains that may limit our ability to descend, is not straightforward. At regular intervals we practice scenarios such as these together in a training center, outfitted with a mock-up of an airplane cabin and cockpit, with discussions afterwards to review the identical technical situation from our different perspectives. The briefings before each trip are a chance to recall those exercises together, as the newly formed team might be faced with such a situation only a few hours later.

Separate from our discussion with the cabin crew, the other pilots and I review the technical details of our flight—the route, any closures or temporary deficiencies at airports we will fly near, any minor problems with the aircraft, known as *acceptable deferred defects*, that we must look up in a voluminous manual.

Weather is an important topic. Our briefing typically starts with a map that displays our entire route. Sometimes, such as for certain northerly routes to Japan, these maps are centered on the North Pole. It takes a moment to decide which way to hold such a piece of paper, because it hardly matters. The charts are scattered with curious, weather-related markings—*meteoro-glyphs*, we might call them. We point out the jet streams and the regions of potential turbulence or storms or icing, drawn to look like clouds though they are often the size of countries. Typhoons and hurricanes are marked with a simple circle with two spinning tails; it resembles the technical icon in our manuals that denotes a pump, which is not an inaccurate analogy. There is a symbol for volcanoes—a pyramid with the top missing, with little worried dashes of lava pouring up from it—and one for radioactivity. A pen traces over this world at hundreds of miles per second. Turbulence here, icing possible there; a spilled drop of coffee; more turbulence there, during the breakfast service tomorrow, unfortunately; storms here; a volcano there.

Next we consider the weather at our destination and the airports near it, accounting not just for the specialist aviation forecasts—written in a code now so familiar to me that I use the forecasts even when checking the weather at home—but also for each pilot's experience, as other pilots may already know well the atmospheric eccentricities of a destination that is new to me. São Paulo, for example, is famous for its heavy rains, which are not always forecast. In San Francisco, the winds often blow more strongly near the ground—a reversal of the sky's usual arrangement of motion, although even forceful winds near the ground there are rarely turbulent. At Narita Airport, near Tokyo, however, even light winds can be surprisingly bumpy. Over the years this rhythm of forecast and experience accumulates to a satisfying awareness of the meteorological personalities of the world's cities, as if they were characters, colleagues we have naturally come to know well.

Flights often use a little more fuel than a straightforward calculation would suggest. The wind forecasts sometimes turn out to be not quite accurate, or the taxi out takes longer, or we cannot fly at our optimum altitudes, or congestion delays our arrival. We are given a statistical allowance for all this, based on a detailed history of fuel usage by flights on that route. Individual planes may also be given a *fuel factor,* an increase in fuel that reflects the consumption history of each aircraft, each individual hull, which can be thought of as similar to a handicap in golf. Outside the briefing room, once our fuel decision is agreed, we enter it into a computer that transmits it immediately. Fueling a long-haul airliner is not a quick process, and by the time we ourselves reach the aircraft, less than an hour before departure, it should be well under way.

One of the milestones of every pilot's career is their first *solo,* the first time they fly without an instructor. Ceremonies and traditions abound; upon landing, a newly soloed pilot may be drenched with a bucket of water or have their shirttails cut off. I soloed near Phoenix, Arizona, in the early part of my flight-training course.

After my first solo, the remaining visual flight training consisted of a mixture of solo flights and flights with an instructor onboard. Near the end of this period of mixed flying my instructor pointed out something I had not realized. When you return to England to start your instrument training, he told me, all of the flying will be done with an instructor onboard. Enjoy your last solo flight, he said as I walked out to the aircraft on a sunny afternoon, because unless you later decide to fly privately, you will never again in your life be alone on an airplane.

He was right. Many small planes can be flown by a single pilot. Yet commercial jetliners do not have two pilots merely for redundancy. Everything about how they are designed and operated assumes the presence of two pilots—a captain and a copilot, more formally known as a first officer. Both first officers and captains are pilots and do roughly the same amount of flying, but the captain—who must sign many documents before and after each flight and, like the postilion of horse-drawn coaches, always sits on the left—has additional managerial responsibilities, and ultimate legal authority, as commander of both the aircraft and its crew.

Most other tasks are divided strictly between the *pilot flying*—whether they are flying manually, or through an autopilot—and the *pilot not flying*, or *pilot monitoring*. Captains and copilots alternate between the pilot-flying and pilot-monitoring roles. One of the least appreciated aspects of airliners is that they are designed to be flown not only by two pilots, but by two pilots swapping between what are essentially different jobs.

The captain decides who is the pilot flying for each leg, a decision that must account for several variables. For example, pilots must perform a number of flights and a number of landings within a specified period, in order to maintain what's called *recency*. Weather is another consideration; as a copilot I am not allowed to be the pilot flying for an automatic landing in fog, for example. So if we are flying from London to New York, and fog is forecast for our return to London in a few days' time, the captain may say to me: "Why don't you take it out? I'll bring it back." Out is New York; back is London; it, the controls of the 747. Longer flights feature an additional copilot, known as the *heavy*. A captain, meeting the two copilots for the day's flight, may ask which one of us is *heavy out* and which is *heavy home*. On some long trips a captain may not be the pilot flying in either direction, because both copilots *need a landing* to maintain their recency. The longest flights have two heavies—a total of four pilots, one pair in dreamland while the other pair flies.

The interaction and cooperation between the two different pilot roles—pilot flying and pilot monitoring—is highly formalized. Not only the tasks but the language that accompanies their completion is specified in exacting detail. When the flying pilot decides to lower the landing gear, for example, they do not reach for the lever themselves. They say: "Gear down." The monitoring pilot repeats the instruction out loud, to check that it has been heard correctly, then checks that the speed and altitude are appropriate for this instruction, and only then actually moves the lever. The division, then, could be likened to that on a road trip with someone you get along well with. Only one of you drives. The other checks and gives directions, changes the music or temperature, passes snacks and drinks, searches in a guidebook or on a smartphone for the best diner in the town ahead, calls the motel to see if they still have a room.

The checklist is an important aspect of aviation safety that has been recognized by other fields in recent years; it has migrated most notably to medicine, where checklists have been shown to dramatically improve compliance with a seemingly simple series of important steps, such as those that reduce infections after the insertion of lines into arteries and veins. But what surprised me most about the use of checklists in the cockpits of airliners is their strictly interactive nature.

When it comes to all the most important checklists on an airliner, it takes two. One to read the checklist items, known as the *challenges,* and one to respond. The flying pilot calls for a checklist by its name—*landing checklist,* for example. Only then will the monitoring pilot remove the checklist from its holder (many planes have electronic checklists now, but the principle is similar), and read its title out loud, followed by the first item, which might be "speedbrakes." The flying pilot will then check that the speedbrakes are armed, and only then reply: "Armed." The monitoring pilot will check that this response is the correct one, and then move on to the next item. When the checklist is complete the monitoring pilot will announce: "Landing checklist complete," and stow it carefully.

The teamwork that flows from such clearly demarcated roles shapes everything about our hours in the cockpit. If I'm the pilot flying and I wish to stand up to stretch my legs, I must turn to the pilot monitoring and say: "You have control." I am no longer driving; you are. Only when they respond: "I have the airplane," or "I have control," may I reach to undo my seat belt.

This formalized cooperation between pilots (similar principles of roles and teams are at work among the flight attendants, and between them and the pilots) is like nothing I have experienced or read about in any other context. Yet this remarkably close working environment, this highly structured mesh of roles and teamwork, takes place among strangers. We fly away together, and are very much alone together, high above the sleeping Arctic or Sahara, and then we are alone, too, in a far-off and foreign city, where we share our jobs and our foreignness, and we may eat and talk together until late in the evening. The next day some of us might meet up to explore a new neighborhood, or rent a car to see the nearby countryside, or join a colleague who has a hobby we are inclined to share. That night, or the next, we cross back over the world together.

When we return home we remove our bags from the baggage belt, smile and shake hands, thank each other for a good trip. The scale of our common journey, to and from the far side of the earth, is breathtaking. What more could bind us? Yet there is a reasonable chance we may never speak to each other again in our lives.

If we do meet again it may be several years later, and it is likely we won't remember where it was we first flew together, or when, or what we talked about in the course of a long evening or two. I don't think I'm alone in knowing the embarrassment of remembering a colleague's face or name but nothing about their life, though there was an evening six months or years earlier when we told each other a great deal about our lives. Often I meet someone, seemingly for the first time, and then a day or two into a trip, over dinner, they tell me some memorable detail of their life—their uncle's health problems, the pet shop their partner owns, their penchant for deep-sea fishing—and I remember that I have met them before. I remember their story but not their face. A large portion of my daily social routine is washed clean by repetition, by volume, by the simple limits of memory.

When I read a scientific article that makes some reference to the typical size of groups of prehistoric humans, the tiny community that would once have been our whole world, it's natural to reflect on the dizzyingly enormous metropolises I visit, and also on the disconcerting ease with which aviation turns from one such city to another. But mainly I think about how as airline crews we meet, work closely together in an environment that demands a heartening and particularly pure sort of ritualized teamwork, and then we say good-bye.

The airline's turning world of briefly encountered faces may appear to be a sad thing to outsiders. Certainly it is not something I would have been drawn to, had I known about it before I started to fly. But I've come to appreciate certain aspects of it.

Among the unforeseen advantages offered by our enormous and anonymous community, one that I like doesn't have a name, really. It is a feeling I associate with the term *face value*. When a crew meets for the first time, we know only two things about each other: that we've each met the standards for our roles and that it is almost time to go. In such circumstances a natural warmth, a practical case, is both a necessity and a regular reward. There's no reason not to want the best from each other; it's goodwill in the simplest sense.

There is another quality to the profession, the same gratification I occasionally found in the paper route I had as a teenager. There were many snowy subzero mornings when I would not have put it in such auspicious terms, but I occasionally felt a small, inverted pride in unrecognition, in working while the rest of the world slept; in knowing that most people would only think of me if I were

late or absent. I imagine that power-plant workers, or snowplow
drivers, know the same quiet pride of starting early, or finishing
late, so that the rest of the world may work.

There's also a camaraderie between the crew and the ground
teams, those responsible for the vast array of tasks that are required
to send an airliner on its way or to welcome it on the far side
of the world. Among all these—the check-in and boarding staff,
the engineers, the caterers and cleaners—there is one person that
pilots work with most closely. The term for this role varies with
the country, the heritage of the airline, and the specific duties—
turnaround manager or coordinator, or *dispatcher*, are often heard.
But in my corner of the industry it's *redcap*, for the distinctive hats
these staff members often wear, that I hear most often.

Among other duties, redcaps are responsible for coordinating
the departure of an aircraft—the baggage loading, the fueling, the
catering, the passenger boarding. Their job is to *turn* the plane, to
get all 370 tons of it pointing back in the other direction as soon as
possible. "I'll get you away on time," a redcap may say, after intro-
ducing themselves and the first half-dozen of the problems they are
in the midst of solving.

Abroad, redcaps link an international organization to teams
of local workers—in this sense they are like the dragomans of
old empires—and far more than pilots they embody the age of
globalization. Many local airport staff may not speak English, so
redcaps speak the local language (or two or three local languages,
in some countries), in addition to English. E-mails and conference
calls connect the world in one sense; but there is nothing virtual
about a 747 landing in Chicago or Accra. The baggage containers must fit; the fuel and the fresh blankets must be waiting. You
cannot redial or resend; many people and things must be ready,

standing in the wind and snow or the bite of the equatorial sun, waiting for the moment the jet will appear from the sky and park before them.

Redcaps are almost always in motion: up to the gate, down to the tarmac, through to the cockpit; within a few short minutes they may speak to the pilots, the flight attendants, the head office—many hours and miles away—the aircraft cleaners, and the caterers without skipping a step or beat. I am not sure I would recognize a redcap if I saw one who was not moving or whose phone was not ringing. For many years we celebrated Thanksgiving with a large group of English friends in London (an opportunity for me as host, not only to share a tradition but to adapt it, by removing the pumpkin pie that I've never liked and few guests knew to miss). But with only a small oven it was always a challenge to coordinate the cooking and reheating of various dishes. One particularly chaotic year, when I was just back from a night flight from Lagos and juggling half a dozen trays and taking a cigarette lighter to the feather stubs of an all-too-authentic English turkey that had not been properly plucked, it occurred to me that redcaps could do a roaring side trade in event planning, and that surely in their own homes every Christmas dinner goes off without a hitch.

Among pilots and flight attendants, in many ways it's the small details of our wayfaring that bind us most tightly. Occasionally I meet a flight attendant or a pilot from my company in another context—while traveling for pleasure, say, or because they know a friend of mine. They understand that I never know what days I will be on my home continent the next month, or that there's a chance I'll be far away on Christmas Day, or airborne late on New Year's Eve, unsure of when or where to hum "Auld Lang Syne."

We may also share a sense that cities are as much a place as a kind of fixed, time-bounded task, with an urban grammar all their own: "My next Cape Town is August," or "I've got a Nairobi next week," or "Are you Singaporing with us?" They understand if I accidentally use the old name for a city, such as Peking, Bombay, or Leningrad, because the three-letter airport codes—PEK, BOM, LED—are how these cities still appear to us on schedules and in nearly all in-house references; or if I speak of Tokyo as Narita, because its code is NRT, and because the small city of Narita, home to Tokyo's out-of-town airport, is how the world's largest conurbation is marked on our shared and peculiar world.

When I drive somewhere with a pilot friend, perhaps to explore the surroundings of a new destination, I joke about *pilot drivers*, how they (like me) may relax only when they know the next event—the next sign or turn or stop, and its approximate distance, and perhaps even the one after that, too. This attitude may come from instrument flight training, in which we are taught to be always thinking, in terms of both time and distance, about what is going to happen next. Or it may be that people with such sensibilities about life, and roads, are more likely to become pilots.

Other rewarding aspects of my job are, surprisingly, the direct result of the ever-changing roster of faces. I once flew with an older captain who asked me, at some quiet moment of our journey together, what my passion was. He meant, what do I love to do—aside from flying, which he would assume all pilots enjoy. It's a good question to pose among a community that is liberated from the traditional constraints of geography and weekends. What, in this whole wide world, do I like best? Hiking, swimming, cooking badly, I replied.

Indeed, if you enjoy the ebb and flow of meandering

conversations—in quiet moments in the sky over South Dakota or Samarkand, or at a breakfast table in Delhi—the window seat in the cockpit opens onto as many lives as it does places. The sense of foreignness around or below us, the peripatetic but communal nature of our jobs, and the pace of the turning earth, naturally lend themselves to storytelling and candor. Many pilots come to airline flying after careers in the military or in fields unrelated to aviation. The backgrounds of cabin crew are even more diverse. When I hear a colleague's stories from a war, or of flying down tightly controlled air corridors to West Berlin, or of growing up above their father's rural pub in Northumberland, or of a childhood in India and the time they met Tenzing Norgay in Darjeeling, or the years they passed installing cell phone systems in Mongolia, or strapping barrels of oil into a float plane and flying them through snow squalls to lonely lakeside camps in the Canadian north, I understand that my job is showing me the world in yet another way.

Such an enormous cohort of colleagues, who come from so many different places and generations and backgrounds, is a resource. This is true on a professional level, in terms of hearing different perspectives on routes, airports, and weather conditions that an individual pilot may experience only occasionally, and it's true on a personal level, for cities, where an airline crew is a living guidebook, one that many frequent air travelers know to consult about their shared destination. Nearly all the jokes which I retell I've heard first from my colleagues; it is easy to picture these as the cultural memes they are, echoing out across the world along all but epidemiological lines of propagation.

There are the small kindnesses, too, of my job—purer, perhaps, because they are so often anonymous. It's common, for example, for each crew to prepare the cockpit for the next; for a crew they may never meet. To reset the radios with the frequencies suitable for the start rather than the end of a flight; to anticipate the sun's movements by carefully placing sun visors around the cockpits that can take hours to cool down; to dial away the last altitude setting for arrival and replace it with the first setting for departure. It's a good start to a lonely night in the sky, to walk into an empty cockpit and find such gestures set on the metallic indifference of our technology.

If the atmosphere can seem antithetical to deep bonds between those who work high up in it, there are also sparks of real connection. Airline crews often fly over holidays or spend one far from home; there is a natural rapport among a crew spending Christmas in Riyadh or New Year's in Istanbul. Once I had to fly from Britain to America as a passenger due to a family emergency. My manager had met the crew at Heathrow before departure to tell them why I was traveling and on this flight the crew—none of whom I had ever met before or since—treated me with the personal warmth I could only expect from my closest friends, as if they knew better than anyone what our jobs might lack on such nights.

Other times we are brought closer together by the blizzards, hurricanes, or floods that can disrupt an airport or a region for several days or more. In a far-off hotel, the temporary home of this most temporary community, nothing binds as tightly as the shared inability to return home.

I was in Cape Town between flights when ash from an Icelandic volcano closed the airspace over much of Europe. My colleagues and I had planned to stay in South Africa for two nights, but it would be ten days before we finally returned home, and even then we would not know we were leaving until a few hours beforehand. During those ten long days we came to joke about, and then imagine, a Europe devoid of aviation. How would we get home? The size of Africa, of the world, has never been so apparent to me as it was that week, when the mechanism that brought us across it was so precipitously withdrawn. We mused aloud about various overland routes across Africa. Whatever happened to the Cape-to-Cairo railway? Or might we ride by motorcycle up the west coast of Africa, and wash up, in our torn and dusty uniforms, as exiles in Casablanca, where we would wait for passage to Europe? Our forlorn 747 was parked patiently at Cape Town's airport. We joked about phoning London to ask for permission to take the cabin crew up for a morning spin over Table Mountain or up along Namibia's Skeleton Coast or perhaps over Victoria Falls.

The other copilot and I got along well. We went out several mornings on a drive, each day to somewhere new in the Western Cape. We talked about flying and about life; greeted each afternoon's news, that the skies of Europe were still shuttered, with a wry smile and a discussion about where we might explore the next day. I haven't flown with him since, but thanks to the far-off whims of North Atlantic volcanism he is as close a friend as any I have made at work. If we meet again in a 747 cockpit or in a restaurant in some distant metropolis in another ash-shrouded time, we will have a lot to reminisce about.

I regularly meet pilots whose families have ties that reach far back in aviation history; a father who was an engineer on the Concorde, a great-uncle who distinguished himself as a pilot in the Second World War, a grandfather who flew for some illustrious and sepia-bound predecessor of our company. Some pilots are married to another pilot or a flight attendant, and they will sometimes travel together on the same flight. I have heard of two brothers in my company who are both pilots, and two sisters. Fathers and sons fly together occasionally; I have recently heard about a captain whose daughter has joined his fleet.

I once flew with a senior captain who keeps a handwritten diary in addition to his professional logbook. When I asked him what he writes in it, he told me he writes about each trip and includes the names of his colleagues and something of their stories. Whether or not he'll ever fly with them again, he said, he does not like to forget them entirely. He'll be able to recall the days and faces and stories of his long career, as few other pilots will. Such a diary is so rare that it is a form of memory for both of us; I will not forget him for it, either.

I've never carried a diary on trips, but for my first few years as a commercial pilot I kept an old-school, cloth-bound paper logbook in which I kept the legally required record of my flights. In this heavy book I noted the dates and times of each flight, the name of the captain, the registration of the aircraft, the airports of departure and arrival, and whether the flight took place at night or during the day.

Late one year, not long before my mother died, I decided to switch to an electronic logbook. It was close to Christmas and I was at home in Massachusetts, so my mother and I would go to a coffee shop we both liked and sit for hours drinking hot chocolate. She would read her book or the local newspaper while I typed the handwritten details of hundreds of old flights into the new computer program on my laptop, a tedious task that made her smile in sympathy whenever she looked over, although I now remember it with an almost unbearable fondness. Often, when I came across the name of a colleague, I would gaze up through the snow falling outside the window, trying to remember the face of the captain I had flown with to Rome or Lisbon or Sofia one autumn evening a few years earlier, or something from the hours of conversation we had shared.

For many pilots their strongest personal bonds with other pilots come not from work but from their training days. In many countries, especially outside the United States, it's common for aspiring airline pilots to complete a residential course together over eighteen months or so, before they perhaps go on to join the same airline. Those on such a course may become friends for life—friends because they spent a year or two training as part of a team that did not change, for a job in which the team always changes.

I am in Hong Kong. It's late on a steamy morning, the day after my arrival. I'm on the Kowloon side of the harbor, sitting at a café with free Internet access, something airline crews learn to sniff out as effectively as any backpacker. I see an online post from a pilot friend, from my flying course: "When you go to flying college, they never explain the odd things you'll one day do, like sit alone late in a Chinese restaurant in Cairo listening to the local band murder soft rock." We have all known the occasional night like this and we post comments in sympathy—perhaps from an Egyptian restaurant in Beijing.

Very rarely I may see a friend from my own training course overseas, when our trips overlap at a busy destination. And most rarely of all, I may actually fly with a friend. Each time it happens—perhaps ten times in total, so far—I have been excited for weeks in advance. If the disadvantages of my job include the lack of a fixed set of colleagues, and the time spent away from friends and family, then it's hard not to smile when I am with a friend at work, as if it were any job, except that we're watching the approach of dawn over the Indian Ocean, or the coastal ranges of Greenland. This friend, this work; there's nothing else to ask of a day.

In my experience it is common for pilots to train in pairs. In a small plane, this allows one pilot to watch both the other trainee and the instructor from the backseat; and in the simulator-based training for airliners that are flown by two pilots, the "flights" run most smoothly when there are two pilots at the controls, one in the flying and one in the monitoring role, while the instructor directs and corrects from behind. Sometimes after a training exercise we watch a video of our interactions with the other pilot, so the instructor can point out, for example, when I asked a leading as opposed to an open-ended question.

When I did my visual flight training, in that small plane in Arizona, I was paired with a pilot who has since become a good friend. Sometimes we did solo flights, alone in the plane. But whenever we flew with an instructor, the other was always in the backseat.

One morning we realized that we both had solo flights scheduled. So we taxied out in sequence, and after takeoff we found each other and flew off in tandem from Phoenix south to Tucson. After a huge breakfast we refueled the planes, too, then took off again in rapid sequence. We flew west, not far at all from each other, a school of just two fish, two friends racing each other across the tawny, hauntingly remote mountainscapes of the Cabeza Prieta refuge. We were headed to Yuma, on the Colorado River near where Arizona, California, and Mexico meet.

As we flew we talked on the radio about the land below, or the barbecue that the other trainees were having that evening, or the movie we would go to see, and this bright connection over the blue—rather than air-traffic control or weather reports—

suddenly seemed the purpose of radios, the reason radios were the first electronics installed on airplanes. Even as I flew I hoped I would not forget this, that early one winter's day a new friend and I flew within sight of each other across the desert and talked about nothing in particular.

It's a day I remember whenever I am flying in an airliner and a friend's voice suddenly joins the same radio frequency I am on. Our world-crossed schedules have brought us to what we could never have planned: to the same room of the sky. Occasionally I hear a friend descending to New York at the end of a flight from London, as I am climbing away from New York, as if we alone were personally charged with the maintenance of some unappreciated equilibrium between the two cities. When a friend appears on frequency, we don't chat, but if it's not busy I may dare a quick hello across this exclusive yet most public medium. I will almost never see their airplane or have any idea where they are. I will probably not even know where they are going. Then one of us will change frequency, leave without farewell, ships in the night.

I am in the cockpit, flying from Vancouver to London. Only minutes after departure the city stops and the mountains rise. Thin veins of light run along the valleys below, as if a broad flat place had been folded and the lights had tumbled down the steep sides into the crease. Even these lights linger only briefly in the climbing terrain, and then a world begins that looks much as if there were no people at all upon it. It is a sense that persists long across the night, across the taiga and the tundra and Greenland and several seas until landfall comes over Scotland. It is one of the loneliest routes.

During one of the routine conversations over the intercom system that connects pilots to flight attendants on a large aircraft, one of the crew tells us that a colleague is weeks from retirement. In the slightly wearying game that will be familiar to flight attendants and pilots on big airplanes everywhere, we start calling the dozen-plus intercom stations dotted around the aircraft until we find her. We suggest that she might like to come to the cockpit for the landing in London and, after breakfast is served, she does. I am the heavy, the extra pilot today, so she and I sit behind the other two pilots as the hedgerows of the Chilterns scroll beneath our wings like the webbed cracks on an aged oil painting, erasing thoughts of yesterday's dusk in Vancouver and the sunset embers on the icy peaks that guard the city we left. I ask if she has been in the cockpit recently for landing. No, she says, not recently. She mentions that she's married to a former 747 pilot and that she hasn't watched a landing from the cockpit since he retired a few years earlier.

I ask if she and her husband were able to fly together often during their paired careers. She nods. We loved Cape Town, Singapore, Hong Kong, she says, with a smile. I think of the 747 we are in, of its eventual retirement and that of her husband, which has already taken place. Of hers soon and of my own some day, too.

It occurs to me sometimes that a working life spent among so many colleagues, the teams that disassemble as cleanly as they formed, might be something people are glad to leave behind at the end of their career. I ask her if her husband misses his work. She answers while looking away, out of the window at the turn of England. Oh yes, she says, he does. He misses the people. I ask her if she means that he misses the many hundreds of fellow pilots he must have met in a long career, or the thousands of cabin crew, or his colleagues on the ground, or the passengers themselves. Oh, he misses all of them. Everyone, she laughs, looking left over Windsor as the runway rises ahead of us and the great wheels lower.

I don't know many pilots from other airlines. On the radio, we rarely speak directly to one another; mostly we interact with controllers, although we do not know them really, either.

A pilot may come to know the voices of the controllers at their home airport, even if it's a busy place. Once I visited the control tower at Heathrow, and I was happy to put faces and names to the voices I had heard for so many years—the voices that to a pilot based there are as recognizable and welcoming a feature of home as Richmond Park or Wraysbury Reservoir, when after so many long hours over foreign places they sail past the cockpit windows of the homeward-descending jet. But I have never learned to recognize the individual voices of the controllers at any other airports or of the controllers who cover the airspace in between.

Sometimes two planes fly the same route at the same time, separated from each other only vertically. Such planes may fly within close sight of each other for half an hour or so, until they are pulled apart by the differences in their speeds and the winds. I have occasionally heard one pilot tell another over the radio that they have taken a photo; they then exchange e-mail addresses. I like pictures of airliners well enough, but a picture of an airliner that I am myself flying, over the Atlantic or Namibia or the Andaman Sea, would be something else entirely, especially precious as a gift from another pilot I will never meet. I still have the photo my friend and training partner took of me in a small plane, when we flew together over southern Arizona.

Often we know we are near other planes because even though we cannot see them, we can hear them talking on a common radio frequency, memorably enumerated 123.45. It's used most often to advise other pilots about turbulence, though sometimes it's used for jokes or to discuss something extraordinary we can all plainly see—a meteor shower, auroras, the striking proximity of Venus and Jupiter in the sky before an eastbound audience of up-all-night pilots crossing the dark waters of the Atlantic. If you have ever asked your crew to find out the result of an election or the score of a game that is in progress, they probably used this frequency. As Internet access spreads to the sky, such requests will be something that only pilots remember from a less-connected world.

I heard a story once, that in the 1970s the British tax authorities briefly embargoed or placed a heavy duty on aircraft radios, thinking they were for entertainment. Sometimes a burst of music is played on 123.45; sometimes you even hear a passage of singing on this frequency, followed by a chorus of don't-give-up-your-day-job jibes.

I am over the North Atlantic, halfway to New York. At this stage in flight my communications panel, my *box*, is typically set to broadcast four separate audio streams into my headset: the shared frequency, 123.45; another frequency reserved for urgent matters not related to sports scores; the voice from the captain's microphone; and the communication line to the cabin crew. It's a cacophony that takes some getting used to. We are in the middle of the tracks, those imposing lines of wind-optimized North Atlantic routes published anew for the westbound and eastbound flights of each day and night above the ocean. The common frequency is mostly quiet. A pilot reports turbulence ahead, but we listen and hear that she is at a different altitude, on a different track.

Suddenly I hear an American accent ask if a certain flight

from another airline is listening to the common frequency. Yes, a French-accented male voice responds a moment later, we are here.

The American pilot explains that his wife and daughter are on the French pilot's plane. He asks if the French pilot could arrange for the crew to find their seats and tell them that he says hello, from not so far away in the sky. It's rare on this frequency to hear anything other than clipped exchanges of aviation terminology, sports scores, and colorful banter. Surely everyone in this region of the sky, every pilot within several hundred miles, is now listening.

The French pilot agrees. But the next voice on the frequency, a few minutes later, is not the French or the American pilot or any other pilot. It is the American pilot's wife. The French pilot has invited her into the cockpit. He has given her a headset and told her that she can speak to her husband, from her airplane to his, though the two planes are not even in sight of each other. The American pilot responds instantly, half laughing, to her voice ringing out to him—and to everyone else over a large circle of the Atlantic Ocean. In his whole life the spheres of home and work will never again meet this way, on a crackling electric bridge in the blue.

Whenever I read a reference to some new software that promises to connect us more easily to one another, I think about how such technology has changed the lives of airline crews, allowing them to stay in touch with home in a way that would have amazed our predecessors. But I'm also pleased by the thought of how airplanes combine a technical modernity with an antiquely physical power to connect. Other connections are little more than metaphors in comparison, mere shadows of the actual motion of one person to the city or table or arms of another, cities or tables or arms where almost always they would rather be.

Airports are by definition emotional places. When I think back to the visits my mother made to London, for example, I may remember us in the British Museum or strolling through Green Park; but what I remember most is seeing her when the doors of the baggage hall at Heathrow opened. When my grandfather died, my father flew ahead of us to Belgium. My brother and I, still only teenagers, followed a few days later. As the two of us boarded the plane at Kennedy airport, on a trip that a week earlier we had no idea we would make, I realized for the first time that someone had meant to our dad what he meant to us.

People fly for many reasons. But the calculus narrows considerably as calendars and circumstance close in upon a specific flight. The plane is a narrow channel between two lakes of place, a bottleneck between the sloshing social randomness of daily life in each of two distant cities. Sometimes this effect is extreme: there's a conference, and half the passengers are computer engineers or physicists or archaeologists; or a large and raucous student group is traveling on perhaps their first-ever flight to a faraway place; or a group of elderly friends is flying out to Venice or Vancouver or Oslo together, to start the same cruise through some marvel of the world. On some routes, royalty feature regularly; on others, celebrities, oil workers, religious pilgrims, or aid workers may appear more often. I did not expect my work to reveal so clearly the circulations of humanity in this age, the spectrum of impulses, ancient and otherwise, that may direct someone today to set course across the planet.

One reason I prefer to work on longer flights is that many passengers seem to share my own sense that such journeys are more momentous. On these flights people's reasons for traveling are usually more compelling, almost by definition, because a longer flight takes more time away from one's life and is typically more expensive. In the terminal, and on the aircraft itself before departure, it's easy to sense the increased gravity of longer journeys, whether in the excitement of honeymooners, or just-retired couples, or even in the demeanor of the most seasoned business travelers, who, like their pilots, seem to draw out the act of settling into their seat in proportion to the number of miles they will spend in it.

Among the many reasons passengers travel, I find the idea of emigration most moving. Perhaps because my father made such a journey from Europe to America, or perhaps because I reversed his great journey with one of my own. On most flights, I imagine, is a passenger who is going to a new country to live; maybe the first of their family, or to join those who have gone ahead. The courses of families, cascading down through the generations, hinge on such decisions, and also—in a small but particular and metallic way—on an airplane that once passed through the history of two places and one family.

Pilots' interactions with passengers are limited compared to those of the cabin crew, though, and so, too, are our understandings of the human weight of the journeys we make together. Pilots of bigger planes are most disadvantaged. Large planes may hold more passengers, but the pilots will probably see fewer of them. On my first flight on the 747 as a pilot I walked onto an empty jet and went upstairs to the cockpit. Three-quarters of a busy hour later the redcap told us that boarding was complete. She took her signed paperwork, shook our hands, and walked out of the cockpit, closing the door behind her. Of the 330 passengers onboard, I had not seen even one of them.

Still, as with the connections I treasure to a few among my thousands of colleagues, there are exceptions. There are passenger visits to the cockpit before or after a flight, and not just by children. If you are interested, there is no reason not to ask. Occasionally pilots might be too busy before a flight, but afterward there is almost always time. Parents often take pictures of their children in one of the pilot's seats, and no parent has yet declined my offer to take a picture of them in the seat, too.

Sometimes I take guests into the flight simulator, which is the only way most nonpilots will ever witness the heart of my job and encounter a cockpit as it looks, sounds, and feels in flight. There is no more appropriate footnote to the simulator's technical wizardry than its ability to also conjure up some of the most personal and memorable connections between passengers and pilots. Meanwhile it is the flight attendants who will interact with so many people, from so many cultures, onboard. When combined with the hours spent in more cities than almost anyone else on earth will visit—more, even, than many pilots, who will be confined to the destinations served by their sole current aircraft type—it is hard to think of a profession that offers a broader view of humanity.

Occasionally a passenger becomes ill on a plane. In such situations, again it is the flight attendants, rather than the pilots, who make one of the deepest possible connections, in an often lifesaving reminder of the early links between nurses and flight attendants. (The Iowa-born Ellen Church, hired as the first female flight attendant in 1930, was a registered nurse, as were many of the first women who followed her, until the demands of the Second World War called many nurses elsewhere.) Pilots are involved in such medical situations only indirectly—flying faster, or calling for advice, or considering the option of landing before the destination. The calls for medical advice go via satellite to a central office where doctors evaluate patients on planes and boats in the most remote locations all around the world; virtual medicine at its most necessary. Occasionally the cabin crew seeks a doctor or nurse from among the passengers. Doctors are frequent travelers; I have never been on a long-haul flight on which we needed a doctor but couldn't find one.

A friend of mine who is an airline captain in the United States told me about his early flying days, when he flew small planes for whomever would pay him. Often he—alone, late at night—was tasked with flying a body, with flying someone home who had died while far away from it. This was in the time when banks always returned cashed personal checks to the person who wrote them, and so sometimes he would fly solo through the night with a cargo of one body and several bags of cashed checks. I remembered this story the first time I flew a plane with human remains listed on our paperwork. The additional, perhaps archetypal, sadness of dying abroad is still somehow present even in an age when someone who does so is likely to be repatriated. We do not have a name, or any other details, and perhaps nothing better symbolizes the connections and disconnections of the modern world, that such an important act should be so anonymous to those charged with it.

Once I was in the cockpit of a flight about to depart, when an official car drove straight up to the aircraft, its lights flashing. The driver brought up to the cockpit what looked like a picnic cooler, containing, he told us, human corneas for transplant. The act was as anonymous as the carriage of human remains. We would never know anything of the donor or of the recipient and our role in the gift was entirely incidental. But since then, whenever I've confronted the idea of organ donation, on a driver's license application or when my parents died, I've considered that flight and the persons for whom the corneas were destined, where they are, and how their sight is. I remember that we carefully strapped the box down in the cockpit, made our best speed for London.

Among the many passengers I carry, occasionally I will know one. To fly a friend or family member feels peculiar when it is time to make announcements, to know that one person in the cabin will hear my voice differently, that one person will hear my voice at all. And, they report afterward, the announcements sound equally curious to them. It is the same when friends see me in my uniform, when they are staying with me and I am about to leave for work or have only just returned. Their eyes skip between the face they know and the visual shorthand of my uniform.

Once on a flight, I realized that a neighbor was among the passengers. She didn't know I was one of the pilots. I went downstairs to say hello. I was surprised to find her in a seat over the Atlantic on a 747, rather than on the staircase of our building. Her expression, too, jumped from a blink of confusion to a smile, as I switched from the identically uniformed pilot she knew nothing of, to the neighbor she had so often cooked for.

I am on a flight to Berlin. It's been a long day; the captain and I have already flown from London to Madrid and back, and now it is night; soon we will start our descent to Tegel Airport, to our hotel, to bed. I make an announcement to the passengers about the fine weather waiting for us, and our arrival time, and the view that passengers on one side of the airplane will be able to enjoy of the city center on this clear night.

A few minutes later, one of the cabin crew calls. A passenger who heard my announcement has told them he knows me. They have forgotten the name he gave, though, and so as we descend toward Berlin I have no idea who this could be.

We land, taxi in, park, the doors of the jet and its cockpit are opened. He comes down the aisle toward me, carrying a bag on his shoulder. I recognize him immediately. He is from my hometown, my high school. I have not seen him in well over a decade. He did not even know I had become a pilot. "I vaguely remember something about you liking airplanes," he laughs. He is in Berlin to visit a friend. We share phone numbers and smile at providence, that we'd run into each other so far down the road, and find that we had traveled the last hours of the journey together.

Occasionally an airline pilot flies an empty plane. Such flights without passengers are routine for cargo aircraft, of course, but that is their purpose. To fly a passenger plane that has no passengers feels unnatural. It occurs rarely, when weather disruption has left an aircraft at the wrong airport or when it needs to be moved to or from a maintenance base, for example. I have flown an empty airliner only a handful of times. Even before departure, the idea that no passengers will join us is discouraging. The redcap may shrug when they meet us on such days. Their work is, of course, much easier without passengers, but they do not appear to like it, either.

Flights with no passengers are often flights with no cabin crew either, and so one of the pilots must help close the door on the empty and silent main deck, before heading upstairs to join their colleagues in the cockpit. Opening or closing an aircraft door safely is not entirely straightforward, and until my first flight on an empty aircraft I had never actually opened or closed a 747 door other than during annual training exercises, practicing with flight attendants on an aircraft mock-up, on a door to nowhere. Takeoff on an empty plane is different, too. The jet feels unnaturally light. The absence of passengers is measured in tens of tons, a rare reminder not only of the size of airliners, but of the physicality, the take-this-up-there mechanics of flight.

On an empty flight it is a pilot who must walk through the cabin to conduct the routine safety checks that are normally performed by the cabin crew. On the 747, this means a long and lonely walk away from my one or two colleagues in the cockpit, downstairs and all the way back, past hundreds of empty seats that may be dressed and ready—magazines, toothbrushes, and headsets laid out—for the passengers that are not there.

I'm on an empty aircraft, flying from San Francisco to London. Among the three pilots I am allocated the first break, and I choose to take it in a comfortable seat in the cabin downstairs, rather than in the cockpit bunk, because I've never had the experience of dozing in the entirely untenanted volume of a 747's passenger cabin. Humming to myself, I prepare a luxuriant bed in the nose of the jet, more a nest really, from the all-but-unlimited supply of blankets and pillows. I think of the vast cells of the cargo holds below me, which are nearly full tonight with the computer and biotechnology equipment and fresh fruit and vegetables that are the fingerprint of the California valleys and industrial parks we overflew on departure. Outside I can see the peaks of the snowcapped Sierra Nevadas streaming past in the gathering dusk. But breaks are short enough without sightseeing, and so I lie down to sleep.

What I hear next is the wake-up call at the end of my break. On a normal flight this would be a chime in the bunk area triggered remotely by the other pilots, a pleasant enough noise that is nevertheless burned into every long-haul pilot's brain as the last thing we want to hear interrupting our dreams. On this empty flight, however, my wake-up call takes the form of a public-address announcement, personalized to me from a colleague in the cockpit, broadcast to the hundreds of empty seats and one lonely pilot who suddenly bolts upright in a corner of the forward cabin.

It takes me much longer than the usual sleepy moment to realize where I am. The plane has been flying toward the night of the north and the east, and so it is dark outside and nearly dark inside as well. Scattered oval pools of cold moonlight spread across the cabin floor and roll gently back and forth over the carpet with the sway of the vessel in the high wind. No curtains are drawn between the cabins, and as I look down the full length of the main deck, only a few splashes of light dot the shadowy abstraction of the aisles.

Another copilot once told me about a flight he made on a large aircraft, undergoing tests, that had no interior features yet— no seats, no galleys, no divisions between cabins or decks. He said that, from inside, you could see the fuselage flex and twist in response to the maneuvers of ordinary flight. There's no reason I would be able to see this tonight, but in the near-darkness it's somehow what I'm looking for as I peer down the full length of the empty plane.

I sit in my pajamas on the cabin floor, contemplating for a moment the white noise of the engines and the uninterrupted length of this ghost ship, this peculiar library of numbered and lettered vacancies that we have made and lifted above the low world, that is even now heaving itself forward toward the Arctic.

The phrase *souls on board* comes to mind, an antiquated term that is still heard in aviation when an air-traffic controller, for example, wishes to know the total number of persons, passengers and crew on an aircraft. Many tens of thousands of passengers and crew have flown on this plane and will fly on it; no one who saw only the map of us, the far-scattered constellation of our present locations on the earth, would ever guess that what we had in common was one airplane. I change out of my pajamas in front of the banks of unshuttered windows, which for once open onto a night no less lonely than that inside the cabin.

I walk upstairs and make my way carefully down the dark aisle of the upper deck. The cockpit door has been open the entire flight—there is no reason to close it tonight—and from the end of the upper-deck cabin the softly glowing cockpit screens are as welcoming as a hearth. I walk past the empty seats and through the open door. The mug of tea my colleagues have made for me is steaming in a cupholder by my seat. As I walk in I say: Guess who? And the captain laughs, because tonight there is no one else in the world it could be.

Night

I'm in the cockpit of an airliner at Heathrow that's about to depart to Budapest. I've been an airline pilot for about a year, flying Airbus jets like this one to cities across Europe. All over the continent, the routes, the alignments of the waiting runways, the hotels where we sleep and the cafés where we meet for breakfast, the Europe-shaped maps formed of such places, are no longer new to me. Yet this flight feels as important as any in my life, as momentous as my first flight in a light aircraft as a teenager, my first solo flight in the skies of Arizona, or my first flight on an airliner, because my dad is onboard.

Or at least, he will be soon. The captain and I are on a *tour*, multiple flights over several days, each of which will end in the evening of a different city. We've been taking turns flying each leg. This is my leg—of course it must be, said the captain when I told him that my dad would be onboard. I've done the walk-around, the flight plan is loaded, our checks are complete, the cargo doors are closed, the pushback crew is below the plane, ready to roll. Nearly all the passengers are onboard. But I haven't seen my dad yet. I have a sudden awareness that, unlike every other occasion in my life that one of us has waited for the other, tonight there is no question of waiting.

It's December, not long before Christmas. My dad has been in England for about a week. A few days ago we went to walk around Cambridge on a dark and frosty not-quite-day and somehow were offered seats at the Carols from King's College concert that would be broadcast on Christmas Eve. We sat in the chapel under the great stained-glass windows, mostly the work of Flemish glaziers, and under another Flemish masterpiece, Rubens's *Adoration of the Magi*. My dad will stay longer in Budapest than I will; then he'll head to Belgium, to Flanders, to visit his siblings and their families.

Suddenly I see him. He's one of the last passengers to step onto the aircraft. He is speaking to one of the crew in the galley. The flight attendant brings him to the cockpit and I introduce him to the captain, one of the most senior in the company at the time, who smiles as my dad takes my picture in front of the controls. I explain a few of the buttons and systems to him, show him the digital map of our route. Though now a naturalized American, he is proud, I think, that I have started my career on a European airliner.

We hear the muffled ka-thump of the main cabin door closing, a starter gun familiar to waiting airline pilots everywhere. I reach for my headset, a little embarrassed that I have to ask my dad to leave the cockpit and go to his seat. I close and bolt the cockpit door. I call the controllers to ask for departure clearance. I speak to the pushback crew below the plane, enacting my side of a formal conversation that's specified, word for word, in our manuals. "Brakes released," I say. "Are we clear to start engines?" I ask, as we begin to move backward. "Clear to start number two," responds the voice from below. The cockpit quiets as airflow is

diverted away to the engines, a silence that gives way to an accelerating hum as the captain lights the engine under the right-hand
wing. The left-hand column of my handwritten logbook records
this moment: "Departure from Heathrow, 19:44."

It's been dark for hours already at this time of year. We taxi out,
enjoying one of the pleasures of Heathrow at night that few other
airports offer, a system of green and red taxiway lights that echoes
the voice instructions of the controllers and visually directs our
path across the airfield. There is no delay when we reach the runway. I set takeoff power. We accclerate and lift away from London,
climbing over the southeast of England, passing Dover and the
Channel Tunnel's long approach roads and vast rail yards. Tun-
nels, of all things, are easy to see at night. A bouquet of light paths
fans out from a point, as the narrowly confined journeys spread
in their newfound freedom on the land. We cross the Channel.
Minutes later we cross the far coast and I realize suddenly that I
am flying my dad over his homeland.

On this clear winter night we pass Ostend, then Bruges, where
he studied. I think of *The Nun's Story*, the Audrey Hepburn movie.
Her character traveled from a convent next door to where my
father lived in Bruges, to the Congo, where he too would later
move. When the director Fred Zinnemann arrived in the Congo,
he picked the choir my father had started in the colony to sing
onscreen. So from behind the camera my father conducted his
choir, the other nuns, and Hepburn herself, and then he did it
all again when Zinnemann noticed that the waving shadows of
my father's hands had fallen on the white habits. Next is Ghent,
on the left. Then it's me—on the right, and privileged with the
near-darkness of the cockpit that renders the night land outside as

bright as any of our computer screens—who sees my dad's small hometown set among the lights of Flanders.

Belgium, for all its light, is gone in a matter of minutes. Soon we're over southern Germany; then we pass near Linz, Vienna, Bratislava, following the Danube across the illuminated tapestry of Europe. I think of Europe so often in terms of its peripheral or coastal lands, but flights like this one remind me that just as I may think of Missouri or Kansas as iconically American, so, too does Europe have its heartlands, the central and inland places where culture and geography each lend much of their weight to the other. Ahead now are the lights of Budapest. We make a languid arc to the south of the city, then turn back to the northwest to make our approach to the easterly of the two parallel runways.

Like London, like Brussels and Vienna, like everywhere we saw tonight, Budapest is cold and clear. Not a breath of wind is sensed by the flight computers as we start the final approach and extend the flaps. I remember that my dad is onboard and I wonder for which one of us this experience is more unexpected. My dad sometimes said he wished he had become a scientist. I have a flicker of sadness about the rules that mean I can't show him this view of the lights that lead to and mark the runway. He would love the way they look: technical but majestic.

A green bar that may edge out sideways, in *wing bars,* marks the beginning of the runway itself—the *threshold,* the liminal rite cast up in light. Before the lights of the runway itself come the approach lights. When it comes to approach lights, there are many schools of thought. Each runway's complicated arrangement is identified on our charts by diagrams and acronyms that barely simplify them. Sometimes a stream of strobe lights races toward the runway—a *running rabbit,* as if airliners were greyhounds on a track. Some approach-light arrays are more than half a mile long and extend out far into open water, where their lanterned purpose appears more likely to be nautical than aerial. Sometimes, in mist or snow at night, particularly if the airport is surrounded by water, then for several minutes the runway lights can be all we see of the approaching world. Their patterns create a glorious visual momentum; long streams point and narrow toward the runway, cut by arrowing crossbars. Precision blooms in the windscreen.

As we descend toward Budapest the plane starts to speak to us. "TWO THOUSAND FIVE HUNDRED," it calls out. We lower the gear. The glittering patterns of lights, the lampposts of the returning world, are no longer only ahead of us; we are among them; they are streaming directly under the nose. By some grand luck, some pleasing and memorable coincidence of air and family, the landing is one of the smoothest I've ever made. We taxi to the gate, read the shutdown checklist. I complete the entry in my logbook: "Arrival in Budapest, 22:02. Dad onboard."

When flying is spoken of in cultural or emotional terms, the sky is almost always light. The loveliest break with this rule—Saint-Exupéry's *Night Flight*—describes a lower sky and a lower-wattage world, wonders that remain accessible to the intrepid pilots of small planes over rural or wild places but rarely to the modern air traveler, for whom it is easy to forget to look out at night. From an airliner the night world is more subtle than the day version, even from a dark cabin, and it's certainly harder to photograph. And passengers who fly at night are often asleep or hoping to be.

But whether as a pilot or a passenger, I much prefer to fly at night. There is a delicacy that's the opposite of the solar glare we must shield ourselves against, with sunglasses and elaborate phalanxes of sunshades that on long daylight flights migrate like the faces of sunflowers around the cockpit. Night flights are often smoother, too, without the sun to raise heat and turbulence from the earth's surface.

The sense that in taking flight we leave behind the small concerns and low ceilings of daily life is markedly stronger at night. In conversation we may speak too negatively of a "dark night of the soul." The poem by St. John of the Cross is not about despair but about a love that we can see more clearly at night, when the navigation light on the wing, rising over the sleeping lands and cities, may recall the "lantern bright" in one version of the poem; and, in another version, the night beauty of journeys that begin "in darkness . . . my house being wrapt in sleep."

In the high night, too, are many phenomena we cannot see so clearly, if we see them at all, when the sun is up. There are nameless ships of cloud that seem to sail best under a bright moon. There are vast lobes of lightning, flashbulbing out from deep within the gray matter of distant equatorial thunderstorms, while on the windowpanes St. Elmo's fire, a kind of static that appears in startling bursts of flat blue veins, flickers like Prufrock's "nerves in patterns on a screen." There are the empty, passing lands directly below us, dark and almost as far from us in our imaginations as the heavens. There are the flames, both man-made and natural, and more than we would ever imagine. And there are the illuminated manuscripts of cities and small places—the book they make of our lights under the dark-fallen hours, as if flight had been granted only to help us remember that there is a grace to the lights we place on the world; to remind us that everything we know is embowered by stars.

*

It was once said that the British Empire spanned so much of the globe that the sun would never set on it. An Indian-born professor of mine in college, when he found out I was moving to Britain, warned that after a few wintry weeks in the heart of the former empire I might find myself wondering whether the sun had ever risen on it. On the ground, sunset is often an unsatisfactory affair, affected or obliterated entirely by clouds, pollution, and weather, and further handicapped by the fact that, unless we are sailors or farmers, we rarely have a clear view both down to and along the horizon. Indeed, on an overcast day there is often no sign that either the earth or the unseen source of its illumination are celestial bodies. The sky gradually darkens in a generalized and directionless fade from damp gray to wet black.

By contrast, in the sky at high altitude, the coming of darkness is almost always pristine. Nearly every sunset I have seen in the sky would make me stop in my tracks if I saw it from the surface of the earth. It is an advantage of the profession an aspiring pilot may not have stopped to consider, that every sunset will be so perfect that we might roll our eyes if we saw its like on a postcard.

Flight also offers us an opportunity to both scramble and unveil the mechanics of our light and our sphere. Darkness comes to an airliner early or late. It may last unnaturally long, or it may come only in part before starting its retreat. Often darkness does not come at all. Night, on the ground, is experienced as time—nighttime, we call it. In the sky, the intrigues of darkness appear more sensible if we imagine night as a space—a geography of shadow that we can race toward or flee from, at speeds fast enough to accelerate the turning of the day or to all but hold the hands of a clock in place.

We might picture what we learned once in school but now may only rarely consider: the earth floating in the light of the sun. Using an apple and a flashlight can help remind us that at every moment the back of the planet is dark and the front is light. The two halves meet in a continuous belt around the earth where day and night are always beginning or ending, a great ring of light-meeting-dark. This ring is sometimes called the *terminator*, but the line is as much a beginning of light as an end. Along the ring it is always dawn or dusk—the names of two of Isak Dinesen's Scotch deerhounds, incidentally, who when they accompanied her on safari would scatter the game like "all the stars of heaven running wild over the sky."

From our earthbound and seemingly stationary perspective, we imagine that this ring moves over the earth, bringing to each place the familiar rolling pattern of light and time—dusk, night, dawn, day, dusk again. But really it's the earth that turns, not the ring. We turn inside it. It's perhaps easiest to picture if we imagine an observer on the sun, looking out at the ring. The ring is a hula hoop, upright on its rim; turned so that its center is open toward us, it always faces the sun directly. An observer on the sun would only ever see the daylight side of the planet. The boundary between day and night, between what such an observer can and can't see, is the ring. The earth is turning inside the ring, from west to east—from left to right if you arbitrarily place the northern hemisphere on top, as nearly all our maps and globes do.

Picture a fixed point on the turning earth, your hometown perhaps. It turns into sunlight—it emerges from the dark side of the planet into day—as it crosses the ring's left edge. Here is your dawn. Your hometown then passes into the naked daylight at the open front of the ring and moves across the front, across the day. Later, it crosses the right side of the ring. Here is dusk.

The ring helps to explain what happens to darkness and light in the air, especially on long-haul flights. When a plane lifts off from earth in daylight, it may move east. Then it is racing toward the dusk edge of the ring, over an earth that is already doing the same. Their easterly speeds add together, and dusk will come quickly to that plane—or, we should say, the plane will come quickly to dusk, to the right edge of the ring. The plane will speed into the darkness— into what we might just as sensibly call night space rather than nighttime—and may even race all the way around, back into dawn. Such are the abbreviated nights we experience on many eastbound flights, from North America to Europe, for example.

If the plane heads west, however, its direction is opposed to the rotation of the earth. So the plane remains on the front of the ring, in the daytime, the day-place. That is the long day we experience on a westbound daylight flight. Meanwhile the city we departed from is turning away, reaching dusk when and where we would have met it, had we not earlier in the day boarded a westbound airliner. In terms of light, this is what it means to fly from Singapore to Dubai, Muscat to Casablanca, Atlanta to Honolulu. We arrive in the heat of our destination's afternoon even as the city we departed from has been dark for hours.

Marilynne Robinson, in *Gilead,* describes our rotation in the night ring, our transit of the sun's perpetual light, in language that every pilot will recognize:

This morning Kansas rolled out of its sleep into a sunlight grandly announced, proclaimed throughout heaven—one

more of the very finite number of days that this old prairie
has been called Kansas, or Iowa. But it has all been one day,
that first day. Light is constant, we just turn over in it.

The truth of this, seen from airplanes, is all but religious in both
weight and simplicity. At many latitudes darkness need never come
to a westbound airplane, for as long as it can fly.

A plane may remain in the day like this—or in the night, or
on the boundary between the two. Sometimes a westbound flight
departs near dusk, the time and place at the edge of the ring. If it had
remained at the airport, night would have fallen on it. But it did take
off and head west, and the brief hour of dusk may now last for the
entire flight; a dusk as long as a day, made at the ending of the day.
Something I see often: to one side the sun, below a white subarctic
landscape lit lipstick-pink by the sear of low sunlight, and across the
sky—I have only to turn my gaze—the curve of night following us
like an apparition.

Such a dusk, though seemingly permanent on some flights, is not
stationary. Leaving London for Vancouver on an autumn afternoon,
we start by heading roughly north, and the sun is already setting to
the west, to our left. In the middle of the flight it may be ahead of
us, on the nose. Then, near the end of the flight, when we approach
Vancouver heading nearly south, the sunset is on our right, having
moved around the horizon as ordinarily as the hand of a clock or
the shadow of a sundial.

The days and nights of flight are scrambled by a further detail.
Imagine again the flashlight and the ring of light and dark it casts
around the apple. Now imagine a pencil passing through the apple
from top to bottom, the eraser where the apple's stem is, the pencil

tip coming out of the bottom. The earth's axis—the pencil, the line from the North Pole to the South Pole along which the earth spins—is tilted. It does not pass through the night ring, the rim of the hula hoop, except at the spring and autumn equinoxes, when the sun is directly over the equator.

For part of the year the eraser of the pencil is inclined toward the sun, as if you tilted the apple so that the eraser leaned out toward the flashlight. This tilt is the source of the seasons and of the changing lengths of days throughout the year. When the top of the apple gets more direct light, and for more hours of each day, it is summer in the northern hemisphere. And because the top of the earth is tilted forward in the ring, the area immediately around it turns around and around without ever crossing the night ring, and here, therefore, for part of the year the sun never goes down. Indeed, this is one formal definition of the Arctic—the portion of the northern world where the sun remains above the horizon for at least one entire twenty-four-hour period in the light course of the year.

Meanwhile, at the bottom of the apple, the pencil tip sticks out the back of the ring. This is winter in the southern hemisphere, and the area of the apple around the pencil tip never turns into the front of the night ring—the perpetual night of the Antarctic. All this reverses later in the year, when the pencil eraser points out the back of the ring and the tip out the front. Summer and light fly south, as simply as 747s, or the Arctic terns that permanently bounce between summers and so may experience less darkness than any other animal.

The tilt of the earth's axis results in further airborne idiosyncrasies. When the Arctic lies in perpetual darkness, then even on a westbound, so-called daylight flight—from London to Los Angeles, for example, which takes off in the afternoon of London and lands in the afternoon of Los Angeles, never once overflying a place where the local time is not afternoon—the great circle of the flight path nevertheless may take us not only into the geographic north but geographic night. We cross the dusk–dawn ring near the top; we fly into the volume of darkness that rests upon the night lands. The sun may set entirely. The stars come out. And then, some time later, as the great circle curves back toward the south, we cross once more into the front of the ring to experience a second dawn to our day. An observer on the ground below, meanwhile, might see something of this same light in the sky and would call it dusk.

Sometimes this new dawn, this new day conjured up by the plane, will last for hours; it may turn into something approximating open daylight. Other times the sun reverses direction and the dawn retreats; after our extra sunrise, another sunset. And then? I have been on flights where the sun has set and then risen three or four times. In ordinary terms—the day as the time between sunrise and sunset—I do not know how many days one day can hold.

On eastbound flights in the northern hemisphere's summer, when the far north is in perpetual light, other violations of solar decency occur. On a so-called overnight flight from Europe to the Far East we head northeast, and the sun moves behind us. It lowers in the sky but it does not set. Then it swings across the sky, from left to right, until it is due *north* of us. We are watching an entire day take place on the other side of the planet, watching it over the top of the world, where it appears as something we might call a north-set, or a north-rise, blessing us with half a dozen hours or more of the golden-hour light so valued by photographers.

On such flights I have seen the red low sun hover over the far side of the earth, and I have mulled over several hours and cups of tea whether the boreal maneuverings of our jet and star are best described as dawn or dusk, and whether the light seen from the north is yesterday or tomorrow. Finally the sun completes its lap and appears at last where it must if it is to grace the morning of our destination—Tokyo, say—roughly in the east.

These light effects are often masked by the blinds in the passenger cabin, whose function is to block light but which also temporarily dam up the airplane's time-scrambling stream of motion. Most passengers want to sleep on an eastbound overnight flight and so the rise of the sun a few hours into their journey—if it ever set in the first place—must be hidden. Sometimes I walk in the passenger cabin and it is almost entirely dark. Nearly every passenger is trying to sleep. When I return to the flight deck, and the door between the cabin and the cockpit opens, the full brightness of the new world tumbles out like tools from a badly packed cupboard, dust swirling in the blade of light that falls onto the cabin floor.

In *Wuthering Heights* Cathy ponders the vertical geography of light. The high Penistone Crags "attracted her notice; especially when the setting sun shone on it and the topmost heights, and the whole extent of landscape besides lay in shadow." Higher up, the horizon recedes, as it does when you climb a tall building. You see more of everything, more of the sky and more of the sun. The day will find you in the sky before dawn arrives on the earth below, and in the evening, dusk will reach you some time after it has already covered the earth. This is why the sky lightens before dawn and why light lingers in the sky after dusk. We may be pleased by the still-glinting wings of an airliner high above us, leaving a contrail soaked in crimson light, while at street level the sun has already set. We see the plane we are not on, bound for a place we are not, in the last light of the day that has already left us.

When we're flying at dusk, perhaps the sun at last slips below the horizon. Then we climb a few thousand feet, and the sun starts to rise again, in such neat concert with the plane that at such moments it seems particularly absurd to think of light as a function of time rather than space. When I flew on routes around Europe, many winter flights started or ended in darkness. On early morning departures from Lyons or Vienna or Paris we would climb rapidly, passing from a dark, frosty runway into the pure, early sunshine of the day that above us was already dawning. Then in the evening, from the glorious light of a sun not quite below the horizon, we would descend to where it had already set, and from there into a further-fallen night.

There are memorable words for the formal gradations of low light. As the sun begins to move below the horizon, first comes *civil twilight.* Next comes *nautical twilight,* which measures out the last light in which the horizon is visible at sea, an important consideration in old-school navigation. Finally comes *astronomical twilight,* when the sky is dark enough for most observations of the heavens to begin. In aviation, various rules hinge on the definition of daylight—for example, the lighting that's required to be working on the plane or on the runway. A typical aviation definition of day is the time "between the beginning of morning civil twilight and the end of evening civil twilight relevant to the local aeronautical airspace."

In the cockpit we have a book filled with pages and pages of tables that tell us when the sun will rise and set across the world. Its sobriety is an antidote to the antics of the sun, and by reading the entries for en-route cities, we can guess what light we will have when we pass near them. There's no obvious grandeur to this book of light, its utilitarian numbers and place names densely set on sheets as thin as newsprint. Yet when I remove it from its cockpit stowage it feels like a future artifact: a book of the days of our cities, a slender volume from the library of the vessels that cross the rising and falling light between them.

One of my father's former colleagues from his time in Brazil still lives in Salvador, on the country's vast northeastern coastline. Eduardo is now in his eighties. As we have known him our entire lives my brother and I call him our uncle; Uncle Eduardo, his Flemish name Brazilianized as my father's was, from Jozef or Jef to José (a version my dad liked so much that he kept it when he moved to America). Once every two or three years Eduardo travels from Salvador to his native Bruges. He tells me that he always chooses his seat carefully on these long night flights, usually opting for the east side of the plane—the right side as he travels north. This is so he can see the dawn and the first landfall over Europe, he explains when asked, as if such imaginative legroom were every traveler's most obvious consideration when choosing a seat.

Eduardo once asked me if what seems to him to be true—that he can see the day coming from almost an hour away—is really the case. He's right. The clarities of the high night and day are so pure that each intrudes early on the other, each giving the other away, like small children who point at one another and laugh in response to a good-natured interrogation. He tells me that he loves to watch dawn's steady migration of color, the expanding new blues that arc like ripples into the starry black. He says, with a smile, that he could watch forever.

When he is asked by the flight attendants to close the blind, so

as not to disturb the other passengers, he instead places a blanket around the window and continues to peer from under it. He's
watching his journey up and across an ocean, his rare trip north,
along the line from his near-equatorial garden of cacao and cin-
namon to a country—his, my father's—where people marvel at a
vocabulary and accent they now find nearly antique. He is watching the sky and the airplane, their beautiful conflation of time and
distance with his time, his distance.

The next time you are in a plane with the sun setting somewhere on the other side of it from you, look toward the sky roughly
opposite from where the sun is setting. The sky right above may
be almost white, but as your eyes approach the horizon opposite
the setting sun, it turns pinkish and then collapses into a fabulous
series of blues—more blues than any terrestrially born language
would ever go to the trouble to name.

Among the wordless blues is a sight I did not know to look
for when I started to fly. If I had known I might have become a
pilot much sooner. There is a wedge of darkness—a much darker
blue—that starts near the horizon from about 90 degrees on either
side of where the sun is setting. An astronomer tells me his mother
called this "the blanket of night," as we see it pulled over the world.
This sublime slice of midnight blue grows as you move your eyes
toward the point on the horizon opposite the point of sunset. This
darkness is the shadow of the earth itself, projected onto a screen
of air. It is sometimes called the *dark segment* and can be seen from
the ground, too, in the right conditions.

There is an early Islamic title, the "Shadow of God on Earth,"
used by Suleiman the Magnificent and the last Shah of Iran. Here,
instead, we see the shadow of earth on the heavens. This shadow—
the same that falls over the moon during a lunar eclipse—is very

slightly curved. Many intimations of the earth's shape are available to us. But aside from during an eclipse, few of us will ever have the chance to look for the roundness of our planet as directly as we do at nearly every dusk and dawn in an airplane, when we only turn to the window and lift the intervening blind.

My mother liked to give me books on the sky when I was younger, of the sort that mix scientific details with artistic images and tales of how various peoples and ages have interpreted the heavens. She might tuck into such a gift a note, or perhaps a copy of the lecture she had recently heard—by a world-class astronomer, for example, who had started their training in a Tibetan Buddhist monastery.

Years later, after my mother's death, after I have become a long-haul pilot and grown accustomed to so many hours under the night skies of far-off places, I take one of these books off the shelf. I have forgotten none of the impressions it left on me, but almost all of the details. I read again about the Pleiades and rainfall among the Indians of French Guiana; the Milky Way and its ties to the ancient route that pilgrims follow to Santiago de Compostela; I remember that once when I flew the Airbus around Europe and found myself with an extra day in Lyon, I took a train to Le Puy-en-Velay and saw cheerful backpackers walking out of the town on a Sunday morning on the first few steps of their pilgrimages. I open the note my mother left in the book. She has written that she thinks I will very much like this book about the sky. I am struck by the year at the top of the note—1992. I had just started college. It was years before I would articulate a clear wish, to her or myself, to become a pilot.

She was certainly unsurprised to hear that I came to like flying

at night even more than in daylight. Of all the stories she liked to
tell me as a child, stories that often used to cause me to roll my
eyes, the one she told most often was about how I grumbled one
bright summer afternoon: "I don't want the sun to shine, I want
the moon to shine" (the cause, or perhaps the effect, of excessive
readings of *Goodnight Moon*). Like most pilots, blue has always been
my favorite color—"but midnight blue!" as I'm told I used to say.
She had a reverence for natural phenomena and cycles, especially
for those, like that of the moon, that bind us to former ages and
sensitivities toward the natural world—realms to which aviation,
however unnatural it is for us to fly, so often returns us.

If you've ever spent the night in a dark place where you are outside for an unexpectedly long time—camping in a desert, perhaps,
or walking on a dark beach—you may have been surprised to see
the moon rise so brightly that you suddenly remember it is capable
of casting shadows. You realize that for months or even years you
have been disconnected from the moon; you have gone about your
evenings without any direct consideration of it. The brightness
of the moon above a cruising airliner is striking. It is more than
enough to read a map by; enough to cast clean-cut shadows across
the interior surfaces of the cockpit.

My mother also liked to give me calendars of the phases of the
moon, though at a certain point we switched, and it was me who
got her such a calendar each Christmas, from a shop in Covent
Garden. I still order a moon chart every year. But it's more a ritual
than a necessity. I cannot think of another job where an awareness of the moon's states, of the turning of months, might come
so naturally and so free of the intercessions of clouds. The moon
and the sun have the same apparent size to us on earth—they

cover similarly sized circles of sky—a pleasing coincidence that will surprise no one who spends many hours flying, like Cupid seen by Oberon in *A Midsummer Night's Dream*, "between the cold moon and the earth."

Clouds, of course, are their own marvel after dark. If you cannot sleep at night on an airplane, choose the music you like best and turn your gaze to the window, perhaps cupping your hands so that your eyes can better adjust to the realm of the night outside. Moonlight is so bright at high altitude that clouds are as clearly divided as the moon itself into dark and light sides. Such a cloudscape, riven with contours and smoothly curving furrows, looks much like the depiction of our brains in scientific pictures: intersecting lobes of white, their very simplicity somehow a reflection of stately but otherwise unimaginable complexity.

I spend many hours in the night above the North Atlantic, when hardly anyone on the plane is awake. I like it best when we are far from land and a full moon is shining down on a deck of quicksilvered clouds that, like some great species of seabird, are born, live, and die without ever crossing the shore of a continent. The clouds on such a night are more luminous than a snowbound field, with greater texture, but without scale; monumental and silent.

Sometimes scattered cumulus clouds appear over the sea in the light of a high moon, as if called up by some nightly parallel to the process by which the afternoon sun summons their day-born siblings. Under a bright moon such clouds cast clear lines of shadow onto the water. Night is no longer the right word for such a time, for this oceanic realm that grows in the water-floored hours when all is quiet on the plane, when those awake on it may survey this divine workshop where new moonlit cities and courtly lands are spun on the water-looms and silently released from open palms, to sail and vanish over the sleeping planet.

Then there are the stars. The heavens from a dark cockpit are a breathtaking proposition and a consolation when the night is moonless. High up, the sky looks three-dimensional. For once, outer space bears at least the idea of fathoms, of depth; a sea composed of distance, shot through with ancient lights.

So many stars are visible on a moonless night that constellations can be more beautiful but also less important, drawn as they are on the sky from the earth's surface 7 miles down, below a turbulent, humid sky through which only a fraction of heaven can penetrate. Amid such a high cacophony of starlight it's easy to lose sight of the old constellations; easy to make your own, while the Milky Way looks for once like what it is, as it would be if all the stars were droplets of water: a drift of cloud drawn across the darkness.

The sky turns, of course, as the plane moves and the earth rotates. Stars and planets rise; they seem to twinkle both more vividly and more slowly when near the horizon. Or they may flash on and off or between entirely distinct shades with a clarity I have never seen on the ground, as pockets of air act as prisms to divide the star's light, and different colors sweep across the dark windows of the cockpit as if from a lighthouse, or as if embodying an urgent message sent across the night to us in a code of interstellar color.

In the old days globes were made and given in pairs, a terrestrial, or earth globe, and a celestial sphere of the heavens. At night from a plane we may see ourselves as we are: sandwiched between the celestial and the terrestrial spheres, the icy ball of stars turning frictionlessly over us, a high mirror to the steady roll of the dark lands and waters and the lights of cities.

*

I once sent a slightly abstract, black-and-white satellite photograph of the earth at night to a friend. It showed sprawling expansions of city lights, connected by tendons of highway and lit river valleys. I was struck that she later referred to this image as "the star picture," since the satellite's camera was pointing down, not up. Occasionally, over the northern, Mediterranean side of the Sinai, between Alexandria and Gaza, I have seen many pure lights in the water, much whiter than the typical lights of ships; you would swear they were stars, if you did not know you were looking down, your gaze lost in the depth of a different firmament.

Often on flights to southern Africa a colleague and I have watched for the rise of the Southern Cross, a constellation that serves direction-seekers in the southern hemisphere much as the North Star, or Polaris, helps those in the north. A senior colleague who had come to airline flying from a career in the Royal Navy taught me how to make use of the constellation to determine our course, how not to be deceived by the False Cross nearby, which is part of the constellation Vela, meaning *sails*, a pleasing name for an assemblage of light that we see from our latter-day ship. I like to check the plane's digital compass against the Southern Cross and to consider which I trust more, the near-perfect reliability of the airplane systems or my own imperfect readings of an older, astronomical arbiter.

Once I read some letters written by Mark Hopkins, who was partly responsible for the completion of the first transcontinental railroad in the United States. He wrote these letters on a ship sailing from New York to San Francisco, all around South America, via Cape Horn, on the sort of sea journey his railroad would relegate to history. Transfixed by the ocean, he wrote to his brother that if he had had such an experience of the sea when he was younger, he might have devoted himself to nautical adventures rather than the "pursuits on land" that would bring him his fame and fortune.

The captain on Hopkins's ship, after determining the latitude and longitude, posted this information "where all may see and enter in their journal." In the cockpit green digits show our longitude and latitude, much as the captain's notice showed Hopkins the last-known position that he then copied onto his letters, an itinerant and star-sighted address as important a detail to him as the date. Hopkins also wrote about the stars at sea, which have "a serenity and bewitching loveliness in these latitudes such as I have never seen on land." He would be amazed by the stars seen from the cockpit; amazed by the airplanes that cross above his railroad and further shrink the continent.

I am flying from London to southern Africa. We are crossing the east-to-west coastline of West Africa and the airports of Accra, Cotonou, and Lagos roll steadily onto our computer screens, while outside the window a corresponding line of light runs across the haze. We sail past, out over the darkness of the Gulf of Guinea. I give the controller a position report—the waypoint we have crossed and the time we crossed it; our altitude; the next waypoint and the estimated time we will pass it; and the position after that. "Roger," says the controller. "Next report the equator."

I feel a shiver of surprise; I still can't quite believe it's part of my job to announce that we've crossed into the skies of the other half of the world. I try to imagine the old days of the ocean liners, when crossing the equator, the first of our grand marks on the sphere, was still understood as momentous, how on deck, sparkling glasses would be raised.

In the cockpit the equator isn't even marked on our screens. To

know when we cross it we often joke about enacting the imprac-
tical test of watching how water goes down the drain in a sink;
or, more scientifically, we can click through several pages on a
computer screen to call up a readout of our current latitude and
longitude. The last number of these readings are always turning,
as steadily as our engines over the earth. I watch for the moment
when the green digits of latitude reach zero and give way to their
southern mirror, when the N turns to S as the countdown from
the North Pole turns to a count-up to the South. I then call the
controller. "Position equator," I say over the static that often mars
transmissions in this part of the world. "Roger roger," replies the
controller. "Good flight," he says. "Good night."

If you look into the night sky from an airplane for more than a few
minutes—from the cockpit or from a window—you may well see a
shooting star. From the cockpit I may see a dozen during a flight,
without particularly looking for them; my eye catches something,
I look, smile, and say to myself: There's another one. I don't even
mention most of them to a colleague; another will be along soon
enough.

When we pass local midnight—when we cross the halfway point
on the far, dark side of the night ring, heading east toward dawn—
the shooting stars grow markedly more numerous, because the
sky above us now faces the direction of the earth's orbital motion
around the sun, sweeping up more meteors that turn to light and
run across the sky like windswept drops of water over the thick
panes of the cockpit windows. I see so many shooting stars that I
find it hard to think of new wishes to cast into the night, and so
I have settled on a standard one that I feel can bear such astral
recycling. It is more a rhetorical flourish now than a wish, a private

reaction like an unspoken, sky-prompted *Bon appétit* or *Gesundheit,* that perhaps every pilot has.

One winter night, before I became a pilot, I sat in a window seat on the left side of the plane for a flight from Chicago to Boston. It was bitterly cold in both cities. Throughout the cabin the other passengers—mostly businesspeople like myself—were working quietly on laptops or flipping through financial newspapers. About halfway through the flight, I looked out of the window and saw—though I'd never seen them before—what could only be the northern lights. I checked with one of the flight attendants, who had seen them, too, and was watching from the window in the forward door. A few minutes later one of the pilots came out to stretch his legs. He told me he was nearing the end of a long career in aviation, and this night's display was the finest he had seen so far south in the world.

I returned to my seat and peered through the smudged plastic pane. This was in the time when computers would play vivid graphical animations to accompany music, and my first thought was that the display resembled nothing so much as such a screen saver. Soon, though, the snowy earth began to resemble an older world, a deep stage rather than a screen, surrounded by layers of thick curtains of shimmering blue-green light, changing and turning only just perceptibly. I had seen pictures of the northern lights before that night but—as with still photographs taken of the earth from airplanes—the pictures miss so much when they miss the motion. The slow transformations in shape and brightness were like those of milk poured into a glass of iced coffee or dye landing in water.

This is how light can move in the night, above and to the north of our traffic jams and laundry baskets and dentist appointments. In the winter darkness the auroras are clouds of light, from beneath which wisps of illumination drift away, like falling rain driven sideways on the wind. Even in summer, on overnight flights when the crown of the world may never get completely dark, the northern lights may dance to the south, in the darkness of lower latitudes; in this way the auroras bleed across the sky, fading above us into twilight. It makes perfect sense to me that couples flock to hotels in Alaska with the aim of conceiving under this auspicious luminance.

That night, though, despite the pilot making an announcement and the crew dimming the lights, most passengers immediately turned their reading lights on and returned to their papers or laptops, with the world-weariness of travelers who have already had a long day and whose next long day is approaching all too quickly. Many did not look at all. Only a few closed their computers and pressed their faces to the window, to watch the solar wind wash over the lines of magnetism pouring from the top of our home planet. Soon the aircraft began its descent over the forests of western New England and the auroras flickered away to nothing.

My first view of the auroras reminded me how much I wanted to become a pilot. Not long after my flight as a passenger under the northern lights I was accepted to an airline training course. I waited until I had passed my medical tests and then I told my colleagues in the consulting company. I don't think anyone who had flown with me on a business trip was surprised. One said she now understood my great affection for aviation-tuned business jargon—"Let's blue-sky this," for example, or "We've got some strong headwinds in Q2," or "Let's take the 30,000-foot view here," or "They've got a good long runway," to describe an excess of the time and breathing room that an investment gives to a new company. (I'm surprised businesspeople haven't adopted *Wilco*, a military and aviation term for *Will comply* that's heard often on the aircraft's radios. "Wilco—I heard your instruction and I will carry it out.")

Another colleague hugged me and joked about how I had always loved to work in the small shared glass-walled room at the top of our building, the one where we were encouraged to go to do our blue-sky thinking; and about how, on even long night flights, I always seemed to have trouble sleeping.

As the years since I became a pilot go by, though, I find that the northern lights have come to represent a challenge I didn't expect. There are fallacies and complacencies in a life in which auroras are routine and I lose count of the shooting stars on a single flight. Sometimes I find it hard to remain interested—in the northern lights, in the ceaseless numbers of meteors or a hundred other phenomena of the sky and earth—because they appear so regularly; because they are routine to pilots, ordinary by definition.

My original excitement returns, at least in part, when I try to share what I see with others. When I see the auroras start to gather I often tell the flight attendants so that they can look from a window near them or come to the cockpit for a wider and clearer view. They almost always do; for many colleagues the northern lights remain the most revered sight in the sky, one that is especially gratifying in the quiet hours of our long and wakeful nights on an otherwise sleeping airplane.

What pilots decide to tell passengers about the view says much about the place of flight in the modern world. Even during daylight flights, there is a reluctance to interrupt rest or movies with an announcement—and, of course, on a wide-body airplane many passengers will not have clear views through a window. Auroras usually appear when passengers are trying to sleep, so we do not generally announce them. Not every passenger would thank us for waking them up. Sometimes, though, if a passenger is awake—a businessperson perhaps, working through the night on a laptop as I sometimes did—I or one of the cabin crew may quietly point to the window, to the surf of light breaking along the sky's northern shores, and afterward the crew member and I may talk about the sight in the galley, as if it was almost new to us again.

I've been flying the 747 for only a few months. I've completed the simulator training and a series of training flights. For my final exam I have just flown to Dulles Airport outside Washington. When we return to Heathrow, the training captain shakes my hand. Welcome, he says, to the 747.

I am now on my first regular flight since my completion of training. I am going to Bahrain and then on to Qatar, two countries I have never been to before. The leg from London to Bahrain, though about twice as long as any flight I ever flew on the Airbus, is one of the shortest flights for the 747. As we pass to the southwest of Istanbul, I realize that these are new skies, a region of the world I've never flown to as a pilot before; and I remember that the last time I flew near here it was as a passenger, returning from Nairobi to London via the Middle East; I had recently left my graduate program, and wasn't yet sure what was next for me.

We near the Lebanese coast, where I see mountains that I had known about from our charts but that I did not ever expect to find snowcapped. The captain tells me there is good skiing there. A new sky, a new world. Soon we are crossing Saudi Arabia. Before I became a pilot I did not grasp the horizontal dimensions of an airliner's descent, that it can take well over 100 miles; that it is common to start descending toward one country while still high over another. Our vertical journey to Bahrain begins long before we see its lights, long before we can see the far coast of Saudi Arabia.

The return flight to London follows a more northerly route. We cross Kuwait, pass within sight of Iran, and then enter Iraqi airspace. We contact controllers whose accents are more American Midwestern than Middle Eastern. At one point I look down and see a pool of green-gray light beneath us, floating in the hazy darkness. The lights of cities and the outlines they form are often marvelously sharp at night; the etched-glass clarity suggests growth patterns that are simultaneously planned yet biological, an evolved, accidental perfection. But tonight a combination of humidity and rising heat occludes everything at a distance, and the lights below have a kind of fuzzed granularity, like television static or snow. It's the opposite of the crystalline air and light of other desert cities, other nights.

A quick check of the charts reveals that this illumination has a name: it is Baghdad. With my colleague keeping an eye on the instruments, I dim the overhead cockpit lights and press my face to the window. This is what I will later report to friends about this trip, about my first ordinary, post-training flight—that I saw the lights of all Baghdad pass by in the night, and then I ate a sandwich.

Many travelers who ask for window seats are fans of what the earth shows of itself—its natural elements such as mountains, coastlines, rivers and the valleys that cradle them. Such views are a reward of flight, and perhaps the best reason to prize flying in daylight. But many geographic details are also visible at night, when their human significance may become clearer.

In the film *Chasing Ice* the photographer James Balog examines the effects of climate change on glaciers, a topic of interest, perhaps, to those who occasionally overfly the iciest parts of the world. Aside from the vivid images of our changing home, I was struck by his particular affection for photography after dark. There's something about observing the world at night, he says, "that places your mind on the surface of a planet . . . out in the middle of a galaxy." Though airliners take us as far from the planet's surface as most of us are likely to get, when I heard this I thought: Yes, this is something close to why I like flying at night. Night flights remind us that we live our lives on the surface of a revolving sphere, a truth that many of us may see most clearly in the hours we excuse ourselves from it.

Some geographic features on a dark land can be seen directly, when waterborne moonlight falls on a land streaked with rivers or dotted with lakes, for example, or when starlight falls on snowcapped peaks. Others can be seen indirectly, through their effect on the lights of mankind. "You can't divorce civilization from nature," continues Balog, who has seen more clearly than most the effect of one upon the other. And the outlines of this relationship are exactly what we see from the sky, after darkness falls, when a populated river valley such as the Nile is often far more distinct than in the day. After sunset the banks of the Nile turn to paired rivers of light, and even under a thin layer of cloud the illuminated edges of the river are diffused but visible and form auriferous, leopard-like patterns in the cloudscape. Civilization casts outlines of both itself and physical geography up into the night, and through the night's clouds we may see rising up what we are blind to in daylight: the cities of Egypt and the lines of its river.

Mountains, meanwhile, can be discerned by the absence of human light, which may flow around an isolated peak as naturally as water divides around a rock in a stream. When the mountains start at a coast, as they do along much of the northern and eastern Mediterranean, the illuminations of villages and roads are compressed into a golden, littoral plait that lies between the line of unseen water and the shadow of sharply rising land. If the lights of coasts like this one were not so severe in their detail and precision we might call such a view impressionistic and recall Cézanne's description of Monet—"only an eye, but my God, what an eye."

Even during the day, when we survey the works of humanity on the earth, the blessing is both how much we can see and how little. The smaller details are summed or lost. Cars become streams, which become abstracted arteries of motion. Less becomes more: houses turn into communities, communities into a city, a vast city refracts into its bones of light, into only the understanding of it. From the plane we see human landscapes the way a neurologist might sketch an outline of the nervous system: drawings of intricacy, networks, pathways, pulses, and flows that know only their part, not the whole they form.

This distillation—that of millions of individual lives and moments into the physical infrastructure that houses them—is greatly heightened at night. Indeed, so often from a plane at night, the truths of human geography are the only things we see, and all of them in light.

What is so important that we choose to light it? Flight raises few questions as often as this one. A fifth of the world's electricity is used for lighting. Each light we see during our long nights awake above the earth is placed there, maintained with intention. The world still has its lamplighters, though we do not call them this any longer, and we think of them less than we did in the smoky cities of the past. The next time you fly over the glowing dendrites of a populated section of our world, try to imagine the plug pulled, the landscape gone dark, with only moonlight reflecting on water or the occasional fire; dark as the earth was until so recently in the history of our species. When we look down from an airplane we see our civilization engraved in light; we confront the new and stately shock of our bioluminescence.

Some cities are so enormous that from your general position above the world their light-identities are unmistakable: here is Chicago or Karachi or Algiers. But smaller cities may mean more, when you realize that one you used to live in is floating in the darkness beneath you, like a ship you once traveled on; or when you sleep through a long flight and awaken, just before landing, to raise the shade on the gathering lights of home.

Other cities pass by namelessly. I remember, as a child, the feeling of being in the backseat of a car late on the night of Christmas Day, heading home, passing through communities that were not mine, or walking through deserted streets late on Christmas Eve, perhaps to enjoy new snow. At such moments there was a peculiar quality to the quiet houses. Even the Christmas decorations hung on such stillness couldn't capture the weight of the holiday that when you are a child permeates everything. Instead I placed that weight onto the empty streets and the silence and the outlines of nearly dark houses. At night, many cities pass like this on the land. What little we see of the lives within them becomes its own kind of weight.

This sense of the night landscape as a shorthand for the human world is echoed in the cockpit. While cities, countries, and continents are entirely absent from the 747's navigation display, only airports are marked clearly, with a blue circle. We see the light of cities on the earth below; while the world the 747 knows is composed only of blue circles drawn against the darkness of the screen. Over much of the world, however, the shape of land can still be discerned indirectly from the patterns airports make on these screens. Britain, cast in the blue rings of its airports, is easily recognizable, as is all of Western Europe. The eastern United States, too, is drawn well enough from the airport rings that form something like the shape of a continent. Lights on the ground below show an observer in a window seat where people live, and where factories make things; blue circles on a pilot's screen show where enough people live, or enough things are made, to warrant an airport big enough to be programmed into a 747.

I often fly over the Democratic Republic of Congo, where my father lived so long ago when it was a Belgian colony. David Van Reybrouck, a Flemish writer, recently completed an enthralling history of the country, which I am sorry my father did not live to read. The author opens his tale with descriptions of approaches to the country by sea and also by air, an arrival that requires no blue circles in the mind. (The book closes, too, with a flight over Congo, "the huge, moss-green broccoli of the equatorial forest, crisscrossed on occasion by a brown river glistening in the sun.")

Congo today has around 80 million inhabitants—more than Britain or France, about twice the number of California. Its area is six times larger than Japan, something like Alaska and Texas combined. And yet Congo has only two blue circles on our screens, neither of which appear on a more restrictive list of airports where we would consider landing a 747 under normal circumstances. Africa, the second-most populous continent, accounts for less than 3 percent of the world's air-passenger traffic. An observer of modern Africa might watch for changes in the lights seen from above or for additions to the constellations of rings on the navigation screens of overflying airliners.

Congo's most useful blue ring stands for the airport of Kinshasa, a city I now and then overfly between London and Cape
Town. There is often little time to look out at Kinshasa though,
which my father knew as Leopoldville, because the view coincides
with a busy section of flying where the aerial regions of several
countries meet. There are often storms here, too, and even when
the night is cloudless the humid, equatorial air is rarely clear. The
city and the country are among the few that now remain to which
my father traveled, but I have not.

When I have had the chance to look out of the window on a
clear night, there has been little to see. The lights of Kinshasa are
shockingly few for a city of this size, and the contrast between
its scattered lights and the dizzying grids of much smaller places
in the rest of the world is jarring. Every night that I fly over the
United States I pass small cities I know nothing of, that I've barely
heard of, that appear brighter than Kinshasa. Even the light we do
see of Kinshasa looks slightly green and wavy, as if it has emerged
from a boundless volume of water that has absorbed or scattered
the energy of the metropolis. The world remains unequal, in light
as in almost everything else, and the lesson of Kinshasa lies open
to the sky, in the dimmed night-whorl of its fingerprint.

Sometimes when I fly to Los Angeles I arrive from the northwest,
from over the shadowed mountains of Malibu, and the city suddenly looms into view like a bowl of phosphorescence gathered
from the surface of the Pacific. Los Angeles, from a clear night
above it, may alone explain why Joan Didion wrote that "the
most beautiful things I had ever seen had all been seen from airplanes." Even when you come to Los Angeles from the east, from
over the land, the deserts are all but empty until you overfly the

last great crescent of the city's mountains. In this sense the city approached from any direction is an island, ablaze between two oceans.

It is hard to fly into Los Angeles and not to feel that its cultural and geographic positions are as well matched as those of Plymouth Rock; that human and physical geography are hardly worth teasing apart here. The western flow of a nation's cultural energy has reached its terminus and, at night, we see how the light has pooled, like the Pacific it meets, against the beaches and mountains.

If, as a pilot, I could listen to music during an approach and landing only once in my career, I would choose a night arrival in Los Angeles. The air above the city is often free from cloud, and there is a sense that it is hiding beyond its mountains, that its location is privileged by its geography, that we must cross over mountains, or all the Pacific, to approach it. My mother's hometown in Pennsylvania is in the coal country, nestled in a small bowl surrounded by dark hills. When I was a child, each time we drove there at night there was a moment when we came over the last crest of the road into the sudden prospects of the lights of the surprisingly densely settled town below. The view ahead then was as bright an image of a city as Los Angeles is to me now, when, like the starling of Richard Wilbur's poem that vaults over "the sill of the world," the 747 I am flying crosses over the San Bernardino mountains and follows the nerves of the gathering freeways to the ocean.

The name of Los Angeles, too, is perhaps the finest of all the world's large cities, melodic and evocative not only of flight but of a metropolis that to many remains dreamlike. Then there is the scale of the city, the visual oxymoron of its sprawling density, so marvelous at night and from altitude, as if only a city with such a name could be so blessed in light. Turning in flight over Los Angeles after sunset is like turning over nowhere else on the earth. To one side is the electrified bloom of an American-edged night, and to the other, where the wing lifts over the ocean, is a window filled with stars.

A night city may appear after crossing not a continent, but a sea. The East Coast of America, with its large and energy-profligate urban agglomerations, offers singular experiences of night-coast arrival. In the cockpit there are tantalizing hints that a coast is approaching. We may switch from long-distance radios to shorter-range, higher-quality frequencies. The needles on the navigation display that show ground-based radio aids will quiver and spin to life as they acquire the first coastal stations. Shortly after, the first lights appear on the horizon.

Here is a 747, coming to the end of its ocean crossing; here it is tacking high in the ice-wind, making good its steady arc toward the light of landfall. I'm certain that the curve of the planet is apparent at these moments, the land's bowed edge rotating toward us like some long, elegant face, a horizon drawn from the glowing serrations of cities.

The pull of such moments is caught up in the historical weight of what we are doing—the physical crossing of the Atlantic, in six hours instead of six weeks, to reach America and its gleaming nets of coastal city light. Partly, too, there must be some ancient sense of relief at the thought—and then, of course, the sight—of land after a long absence over open water. But mostly I love the visual pace of an approaching coastline. It's something about the way in which the horizontal lights, compacted into a distant line by the angle of vision, gradually expand and turn, revealing the liquid of electricity that we have channeled along the open canals of roads, to the cities that have amassed along the circumference of a continent.

Few coasts are as marvelously straight or well lit as Florida's eastern seaboard appears to be. The boundary between water and light forms a long blade of incandescence that we cross in the last stages of the descent to Miami, all the more striking after six or seven hours of unbroken oceanic murk. For several years my schedule did not take me to Miami, and when I went back at last it was easy to see the city's ever more Hong Kong–like skyline as a kind of jewel, suspended in the night between Manhattan and Rio on a hemispheric arc of cultural longitude.

If I like a song or two about leaving New York, my preferred aerial song of the city would be one of arrival from far out at sea. The city looks as if a huge vase of pixels had been tipped over Manhattan, stacking and tumbling outward, flattening into the suburbs and gradually disappearing into the dark forests of the continent's interior, as if in some computer-age myth of its foundation. The city's bays and rivers glow in this reflected, electric gold; while further out the waters are themselves scattered with the constellated lights of vessels, as if an autumn storm had blown particles of light from the land where they first fell, onto the pitch-dark waters of the city's maritime approaches.

On an eastbound flight from America, Ireland often appears as both the night and the journey are nearing their ends. Even after the most routine Atlantic crossing the sight can bring to mind the "dawn of their return," the words for the hope taken from the eyes of Odysseus' shipmates. The newly sighted land is criss-crossed with creases of illumination, the weave laid over a darkness that looks like history itself. The lights lie most densely around the coast and remain clearly lit even as the horizon above them whitens with dawn. It is a coast that makes me think of half-awoken villages and fishermen already started out from their settlements along the Rorschach-like fractals of the shore. Here is the dawn of return, as simple a view as we will ever have of it; here are lands braided softly with light, and the end of our night's journey across the ocean they face.

Following my father's death—a year and a half after our flight together to Budapest—the world I saw from airplanes, particularly the world I saw at night, changed. Like many people who lose a parent, especially at a relatively young age, I felt that something about the finite nature of life, previously irrelevant or obscured to me, had suddenly come into focus. A nurse might feel time's new weight in piled stacks of medical records, a mechanic in rust and repairs, and an architect in an often-renovated old building's palimpsest of styles. I saw it in what I was spending a large part of my hours above: the human geography of light on the earth.

The patterns we perceive from above—of country lanes and suburban cul-de-sacs, seething freeways, the warehouses storing up whatever it is we will buy tomorrow, the vast pages of parking lots, the steady proud pulse of red on radio masts—is necessarily disconnected from any individual life. We see, instead, the collective infrastructure of all our individual lives, the luminous netting that stands for us but is not us; the lights that suggest a line from Leonard Cohen—"we are so lightly here." If in a moment everyone vanished from a city at night, for some time it would look much the same. The beauty of a night view of a city, though a city is something made of life, made for life, thus has a kind of distance and fragility, a formal or distracted indifference, like the blink-ing language of lit windows on an apartment building as evening draws on.

Above the world at night I was struck by the thought that this—distant, cold, busy, unaware of those gazing down—is how the world might look to my father now. Indeed, a common corollary of grief—the bewilderment of other people going on with their business, shopping, driving, walking, laughing—was heightened, perhaps, because I saw more evidence of it than most.

Friends, unprompted, occasionally tell me about a memorable flight—a flight on which they stared from the window for hours, perhaps, in silence or listening to music, caught by something they had not seen or noticed before. I'm struck by how often the flight they describe to me is one they were taking because a loved one had suddenly fallen ill or died. Such journeys seem conducive to a kind of outward-looking introspection, perhaps because we are likely to be tired or jet-lagged, and because in the rush of calls with family, friends, and doctors, the hours on the plane may be the only time for several weeks that we are alone with our thoughts. We are crossing, too, both mentally and physically, from the time and place before we had this news to some new reality. In the first months after my father died I often wondered how many passen-gers on the plane I was piloting were traveling because someone had died or was gravely ill, and how the lights of the world below the plane might look to them on such a night.

Astronauts have reported that Belgium is easy to spot; on pho-tographs of the earth at night, the country is a continuous splash of white light, as bright as any city. One of Europe's most densely populated countries, Belgium also has one of the world's densest and best-lit road networks. From the altitude of a plane, the lights appear not white but yellow-orange. As I see it so often when flying from London, Belgium appears first as a flat sea of illuminations beyond the shadowy contours of the Channel, a land as densely webbed and light-fractured as a cracked sheet of safety glass, that tilts up toward us in two senses, as we simultaneously fly closer to it and climb higher above it.

Belgium's immediate neighbors survive with less profligate road-lighting policies. This means that on a clear night the sinuous and oft-ignored borders of Belgium are apparent to an aerial observer. The land grows darker beyond their line. I look for the lights of the French city of Lille (or Rijsel, as the once-Flemish city is still known in Dutch), and then let my eye cross northeast, over the frontier of light. This is how I spot my father's hometown from an airplane; how I found it the night he was sitting in the front of the cabin, not 10 feet away from the locked cockpit of the airplane I was flying.

There are other borders that are visible in light; the line between India and Pakistan is one of the brightest and most famous of these light-drawn frontiers. But the sight of my father's homeland marked out in light was dear to me for a long time after he died. So much of someone is where they are from; and my father's past was a different country in more than the usual ways. He told me once how curious it was to return to Belgium and not know the Dutch words for technologies, for example, that were invented or became popular only after he left. In the months after he died, when I climbed out from London and saw Belgium turning toward me on the night eye of the planet, I thought of what lights a pilot in 1931, the year he was born, would have seen from where I was in the sky, and about my aunts and uncles and many cousins, their ordinary evenings passing in the lights ahead of the climbing airplane on this night. Belgium, the land that was most on my mind then, lay before me as the light of memory on the darkness of the past, and the borders of these thoughts were so clear, almost as if, in the nights after a parent dies, everyone's ancestral lands briefly glowed more brightly.

Above the loneliest places on the earth, I've come to appreciate another, less personal experience of light below. Typically, over the nearly uninhabited portions of the globe—the Sahara, Siberia, most of Canada and Australia—you will see no lights at all, or a handful at most. But sometimes, in a very remote area, or if clouds are obscuring adjacent parts of the land, you might see only a single light. A vast sea of darkness—indeed, this effect occurs over the ocean, too, when the airplane overflies a ship—and, floating on it, a solitary light.

A solitary light reminds us of something primal: embers, a beacon. From the airplane all we can see, aside from the light, is the scale of the night that surrounds it; in fact, we see the immensity of such an enclosing darkness far more clearly than anyone on the ground could. A lonely light suggests a fragility and intimacy that cities do not, however beautiful and intricate their nights may appear from above.

When I see such a light, I think back to bitterly cold nights in my childhood, when, crunching through the snow, I would carry a handful of logs back to our house glowing from the soft light of the woodstove. The impossibility of knowing who is down there becomes its own wonder. Is it perhaps a generator in a small village powering a light that will soon be turned off? Or is it several lights—bulbs strung together between houses around a dusty square—that appear as one from so far above? From this high up it is likely to be the numerous lights of a small settlement that have been merged together by our altitude. Will someone whose lit evening has called down my eye look up and see a jet blinking across the stars? Will they speculate on where we are going, which two distant cities our racing light connects? And how long would it take me to find them again, on the ground? Days, certainly; a flight, probably two, to a place with a great many lights; followed by a long, surely arduous overland journey to the place that appeared as only one.

Alexander Graham Bell once prophesied that planes would lift away from the ground carrying the weight of 1,000 bricks. The weight of 1,000 bricks is a little over 2 tons. On the 747, a typical *pantry weight*—an allowance that accounts not for passengers or baggage but merely for the food, drink, and related supplies carried onboard—is over 6 tons, or several thousand bricks (a 747's typical *payload*—passengers, baggage, cargo—is 30 to 40 tons). Weight is a constant consideration in the design process of an aircraft. An engineer on the original 747 cried when the new plane was put on a diet and some of his beloved features were removed.

The weight of an airplane changes dramatically during flight, as fuel is burned. The 15 gallons in a typical car's fuel tank weigh around 90 pounds, roughly one-fortieth of the car's total weight. A jet that departs from Singapore to London may weigh 380 tons. About a tenth of that may be useful payload, while more than 150 tons, or two-fifths, is fuel, nearly all of which will be gone before landing.

The translation of weight lifted to fuel consumed is meticulous. If you carry five extra books in your suitcase on a long flight, an additional amount of fuel, by some calculations roughly equiva-

lent to the weight of one or two of the five books, will have to
be burned to lift your reading matter across the world. Occasionally the number of passengers or the amount of cargo on a flight
increases at the last minute. Then we may have to upload additional tons of fuel to cater to the increased payload.

This vicious cycle results even when the extra weight is itself
fuel. Occasionally—perhaps because fog or snowfall are forecast
at our destination—we load more fuel than the flight plan and our
normal reserves would call for, to allow us to absorb the anticipated arrival delays. On a typical long-haul route, if we wish to
allow for an additional thirty minutes' holding time at our destination, we might load about forty minutes of fuel, to account for the
significant portion of the extra fuel that will be burned merely to
carry the remainder to the far side of the world. And the longer
the flight, the more fuel is burned in this way, which means that
at a certain point the fuel efficiencies of one long flight over two
shorter ones start to fade.

Before I became a pilot, I never imagined that an awareness
of the aircraft's ever-changing weight would become as intuitive
a part of the journey as the remaining minutes or miles of flight.
This sense of the jet's weight is an all but continuous reminder
of the mechanics of flight; of the physical task of lifting people
and cargo away from the ground and across the sky. The aircraft's
weight affects the altitudes and speeds we cruise at, and in certain circumstances even the angles we can bank to in a turn. Our
weight is particularly important to consider when we calculate
our landing speed. We tend to think of heavier objects as slower,
but the amount of lift a wing creates depends on its speed, so a
heavier plane must generally move faster. In the 747 every additional 3 tons of weight at landing—cargo, passengers, unburned

fuel—demand around an extra knot of airspeed. If there are arrival delays and a jet enters a holding pattern, its weight will continue to fall and we must revise the landing speed downward, knot by knot, as the pounds melt away.

Pilots may think of fuel in different ways at different times—as pure weight; as emissions to be avoided; as the equivalent amount of cargo that it displaces, the payload that cannot be carried on long flights against strong headwinds; as time in the air; as miles on the ground; as money to be saved. Amid such practical considerations I occasionally forget that the fuel itself is ancient, and that if something atavistic courses through the clean, technical wizardry of modern flight—if anything can undercut the Whiggish sense that the future arrived with the airplane—it is fuel. The steadiness with which it disappears during flight, neatly exchanged for hours and miles, seems to proceed according to an immovable calculus, first encountered around campfires and oil lamps, that relates certain substances to the quantity of warmth or light or motion our ancestors learned to summon from them.

Sometimes at night this relationship appears vividly on the earth below, in fire, and we might remember the etymology of *petroleum,* rock-oil, and what it means to extract this liquefied power from the apparent solidity of the planet. When oil comes out of a well, natural gas often comes with it. This gas can be captured and sold, but to do so requires additional equipment and investment. For this reason the gas, especially at remote fields, is simply burned off into the sky, or flared.

I'm flying over Iraq, shortly about to start the descent for Kuwait. On the darkness below me stand what look like enormous candles, not votives but tall tapers that appear to be set deep in the desert. Each flame is so bright that it illuminates a perfect halo or sphere of night around itself, like a bubble blown in light, like a round bulb on a fixture that hangs over the porch of a farmhouse surrounded only by dark fields. The light of these fires is red, occasionally golden, although I often can't tell whether the gold is only the color of the flame or of the surrounding land as well.

Such flares are a common sight in the skies over much of the petroleum-rich world. The largest of them flicker or pulse before our eyes. At times I can see dozens of them. They form a new landscape of fire, a terrain marked by torching spires, burning up in the nights of the desert or the subarctic. I associate them with the Persian Gulf, but you also see them over Russia and parts of Africa. When I first flew over Indonesia, en route from Singapore to Sydney, I was stunned by the sight of such flares, embedded not in sand but rising from the ink of the open ocean, from the many rigs there.

These flares have an eerie, even allegorical quality. The wells they crown produce the same fuel the plane may burn; in this sense the fires form a kind of fire-shadow of the plane, and of industrial civilization itself, on the earth night below. They hint at the powers we have unleashed or assumed, to so casually pump fire up into the night. Wherever in the world I see these flares, the sight reminds me of Centralia, Pennsylvania, only a few miles away from my mother's home borough in the coal country, where a barely underground mine fire has been burning for more than fifty years, where steam and smoke rise from cracks behind hilltop cemeteries and snow melts so soon after it lands on the abandoned streets; or of the statue of Prometheus—*forethought*—that stands above the skating rink in Rockefeller Center in New York, under the words of Aeschylus: "Prometheus, teacher in every art, brought the fire that hath proved to mortals a means to mighty ends." Mighty ends, we might imagine, like 747s that cross the stars high above our fire mines, at four times the elevation of Mount Olympus.

We see another species of fire on the earth. Forest fires are a regular sight from airplanes, one that will perhaps grow only more common due to the effects of climate change. Sometimes on daytime flights over the crumpled mountainscapes of the American West, we can spot thick plumes of gray smoke rising from incinerating mountainsides and, once aloft, twisting and churning in the high wind. Winds often change their speed and direction with altitude; the plumes of smoke from forest fires thus act as a graph, a vertical index of the varying winds they rise through.

The return legs of these flights often take place at night, when the smoke may not be visible but the flames themselves sometimes are. The intensity of the fire's brightness and color, distilled through distance, is then chilling and unforgettable, as the wattage of nearly everything else you see is dialed so sharply down. The brightness of such flames is like something from a forge, poured in tiny, molten, fingernail-shaped crescents along slopes that are themselves only shadows. It is as striking a sight as blood on snow.

The heat of a fire causes air to rise rapidly above it, which may then form a cloud known as a *pyrocumulus*. Ice can sometimes form in this fire-born apparition. At other times, like some anti-phoenix, rain may fall from such a cloud and extinguish the fire that created it; or bolts of lightning from it may start new fires, and beget new fire-clouds. In some parts of the world—Cyprus, for example—our flight paperwork may contain a message that asks us to report any fires we see, requests that echo one of the earliest uses of the airplane. I don't think I have ever heard one of these reports given, or even heard *fire* spoken on the radio. If I were going to report a fire on the ground to a controller, I would choose my words carefully.

I'm en route from London to Johannesburg. This is my *hot-and-high* training: a flight to a high-elevation airport in a warm climate, a combination of conditions that presents pilots with a number of challenges.

For now, though, I'm enjoying the peculiar atmosphere of a long overnight flight, the steadier, lower key of this stage of a journey. The passengers have dined and are asleep, the cabin lights are dimmed; one pilot is resting in the bunk, while in the cockpit the conversation between the remaining pilots has dwindled. These are among the night sky's most inviting moments—their quiet, dark, all-is-well quality cradles the vessel, even as we are moving so freely and in utter solitude between the crowded cities at either end of our journey. I like to think of how the plane might look from the outside, in the eye of some balletic, winged observer soaring without effort toward it in the African night. Mostly dark, the blinds drawn, a few running lights on the extremities, the body of the structure outlined against the stars.

Now we are over Zambia, and I see a dull glow on the horizon. In the extreme clarity of night, light often appears from such an enormous distance that it can remain unclear for a long time whether a distant illumination is a city, the rising moon, a glimpse of the day that is unfurling on the other side of the world, or merely the approach of dawn.

A quarter of an hour later, we see what has lit the sky from such a distance: a vast series of fires on the dark plains below us. These are not the brief rims of flame from a mountain forest fire. These are dozens of interconnected curves of light, their shape and visual effect like flame-capped letters or waves on an inky black. Soon we are directly over these blazing runes. They run from under us to the horizon.

I've never seen anything quite so disconcerting from the cockpit. The sight recalls all our older myths and fears, some Jungian-caliber archetype of a fire-rain apocalypse that fractures human order, scatters the animals. The fear of fire is so primal that what's most jarring is our stately presence and progress above such a scene—our *survey* of it, in the precise sense of that word. I'm struck by the contradiction between some deeply innate reaction to such a vision of conflagration and the plane's perfectly safe physical and imaginative distance from it—the cup of tea in my hand, the 300 passengers drifting in and out of sleep, the breakfast trays waiting in their trollies, all the engineering and maintenance expertise devoted to ensuring our safety, from fire above all. The burning land turns toward us and then passes behind us, consumed, like everything else in the world, by the steady motion of the wing.

Soon after the sun comes up we start our descent into a bright, bone-dry, ordinary weekday in Johannesburg, the freeways twisting over the mile-high land in the clear light of morning, the hour humming so steadily that a passenger who'd never been here before could be forgiven for thinking we were descending to Los Angeles. Later, at our hotel, I look for stories about the fires, but find none. I can't believe that something so extraordinary has not made the news. I check the computer again a few days later; still there's nothing at all about the lands we saw burning up in the night.

Before GPS, the small but inherent inaccuracies of navigation meant that planes on the same route would be naturally separated. But once navigation systems were augmented by GPS, the paths of planes could overlay each other almost exactly—so exactly that in some parts of the world pilots reintroduce an arbitrarily chosen *offset* and fly just parallel to the published route, partly to avoid the turbulence of another aircraft's wake and partly to bring back some of the randomness that, along with air-traffic controllers and various onboard systems, forms yet another layer of separation between nearby planes.

Aircraft at different altitudes often cross paths at oblique angles, and the straight fast lines of their motion enact a beautiful precision and complexity, as if sketching out the solution to some word problem in a class on heavenly geometry. At other times, traveling in opposite directions on the same route, planes will pass one another head-on, one right over the other. This is common over Russia and Africa, where long, transcontinental airways are often used by traffic going in both directions.

Though we usually see an approaching plane on our computer screens long before we can see it with our own eyes, looking out of the window remains an important, amazing part of my job. The closing speed of two aircraft is searing, easily 1,200 miles per hour. No sooner do you see the oncoming plane than it is above you, and then it is gone; the fastest event I'll ever see with my own eyes, and a particularly clear view of "this whole new business of speed," as in Faulkner's description of aviation.

I once drove through rural South Africa. After dark the roads were all but empty, and later we were told that it was foolish of us to drive on those desolate roads, far from towns, late at night. On the rare occasions that we saw the lights of another car in the far distance, we would remark on it and mentally brace ourselves for the noise, the bright headlights, the reciprocal, high-fiving gusts that would rock both cars at the moment we passed each other in the night. Then we would talk of something else, listen to several songs, and have an entire conversation before we saw any lights again, perhaps as we came to the crest of a hill. Only after a moment would we realize that it was the same lights of the same car, still far away, which we had forgotten about. The lights at such a pronounced distance in an utterly dark landscape always gave the illusion that the car was much closer.

At night, in a plane, this effect is magnified. We see an approaching plane—or its lights, at least—across a much greater distance than is possible during the day, 40 miles or more, though still, of course, a distance measured in minutes. I like to think of the name of a town 40 miles from home, and what the possibility of light from such unmoored distances says about the clarity of the air and the perfection of the high darkness.

Sometimes on a bus I see the driver wave casually to the driver of another bus passing in the opposite direction, and I don't know how well they know each other, if they are friends or if they are merely strangers exchanging a professional courtesy. Planes always fly with some lights illuminated, but in the lonelier skies of Africa, in the heart of the small hours of a long overnight flight, the pilot of one aircraft, seeing the approach of another, may briefly flash their landing lights. It's a kind of wave across the icy nothing above the jungles or deserts, perhaps under the rising Southern Cross. The other pilot, seeing the lights of the other aircraft illuminate, will generally reciprocate.

Sometimes what appear to be such flashes from a plane at a great distance may in fact be a star or planet scintillating through a long slice of the atmosphere. This effect gives rise to stories of strange sightings and the old pilot's tale of the captain who says to his copilot, who has just flashed the landing lights at what he thinks is an approaching aircraft: Got friends on Venus, have you, son?

I see the reciprocity of the landing lights, this warm yet lonely gesture, most often on flights from London to Cape Town, when we may encounter an opposite-directioned aircraft from our airline, as if each airplane had taken to the sky only to serve as a milepost of light to the other. The term *company ship* evokes the old Union-Castle liners that would leave Southampton and the Cape on the same day each week and pass each other at sea.

High above those waters, while nearly everyone else onboard sleeps, I see the lights of our company ship. I reach up for the pair of landing-light switches on the overhead panels, and the beams from the wings, though only pinpricks in the expanse of the night, salute the opposite vessel as its lights race up and over our windshield. Nothing is said on the radio, and the greeting takes place so quickly I do not even move my hand from the switches. The wings go dark again under the ice-field of African stars and our two jets arc silently past, each bound for morning in the city the other has left.

Return

I'm in the center of Tokyo. The trip into town from our hotel in Narita is a tough one to make in daylight, a swim against a strong tide of jet lag, but it means that my colleagues and I will see something of the city that brought us here and that we will all sleep well tonight. We are walking to a few sights I remember from my short stay in Tokyo in school, when I was on my way to my summer homestay in Kanazawa, and from my business trips to the city later.

We enter the vast plaza of the Tokyo Metropolitan Government building, under the twin skyscrapers that opened a few months before my first visit. The group of fellow pupils I'd joined for that summer adventure was supervised by a twenty-something graduate student from California. She wanted to open our eyes to the world around us. "Wherever you go, there you are," was her favored expression. It struck me then, when the journey was from rural Massachusetts to a country I never thought I'd visit. Now, a decade and a half later, I smile when I think of her words, which articulate the problem of place lag and serve as a kind of incantation against it.

My colleagues and I flew a 747 here this morning, and now we are checking our maps, laughing, looking for a place for lunch. Two days from now we will be heading northwest over Siberia; heading home.

Under the midday light of this distant city, walking and laughing with people I did not know yesterday, I try to picture my slower self, the one whose Stone Age pace and horizons we are all born with. *He* hasn't left London yet; he is puttering around at home. He doesn't even have a passport; he won't fly, ever. He can't conceive of any distance he has not walked himself or seen across an open field or valley. When I get home in less than two days I will meet him, perhaps as he is saying farewell to the apartment; he will be wearing his backpack full of everything he thinks he might need, his sturdiest shoes. I will meet him on the landing outside as I come up the steps. "You don't need any of that. I'm back now," I'll say, and I'll walk past him and take my headphones out of my ears, throw my passport onto a shelf, turn the radio on, put my feet up on the couch, and flip through the mail.

My colleagues and I wander down a side street in Tokyo. We find a small restaurant and have some deep-fried dumplings for lunch. We head back into the sun and I attempt to ask a passerby for directions to the Meiji Shrine.

On the flight from London I asked one of the cabin crew to help me make an announcement in Japanese. I've had to write this out longhand. My language skills have faded in the years since I studied here, to the point that the outlines of the sentences I can no longer form on my own are little more than a phantom linguistic limb. When I first started to fly to Japan as a pilot I was saddened to discover the extent of this loss. I reassured myself that with just a few weeks here I would surely recover nearly all that I'd lost of the Japanese language; I could feel it beginning to return, even in the first day or two. But now I know I won't ever have enough time here to find those words and characters again, at least not with work. I know that a few weeks from now I will have returned to London, then traveled to São Paulo and on to Delhi, where I'll have heard other languages, on other streets.

The language my job has given me—words and names relating to the plane itself, the new geographies of the sky, and the small or far places on the earth I never before knew of—does not replace this. But it has its moments. There is, too, a kind of sign language in aviation which, in the absence of voice contact, allows someone on the ground to indicate to pilots of a moving plane to stop, go straight ahead, turn left or right, or enables a pilot to indicate to ground staff that the brakes are set or that an engine is about to start. These gestures, which echo the semaphores of flags and paddles used to communicate between ships, are internationally standardized. They are drawn and diagrammed with arrows in our manuals to show exactly how the hands must move, on pages that remind me of the sign-language instruction materials my mother used in her speech-therapy practice.

Other gestures are not written out. When a plane departs, a member of the ground staff often gives us a thumbs-up or waves to us, and such a moment is as good an opportunity as any to uncouple words like *farewell*, or to linger on the *God-be-with-you* etymology of *good-bye*. Often in Japan the ground staff, safely distant from the 747, face the jet and bow to it as we start our journey home.

We reach the Meiji Shrine and walk through the regal wooden gate, to which visitors bow as they enter. I have always liked the ceremonial gates that mark a transition between a city, a temple, or a castle, and whatever lies beyond their borders. On one of my trips to Japan, a German resident in Tokyo listed for me some German words that have traveled to Japanese, including *arubaito,* meaning part-time work, from *Arbeit,* and *Enerugii,* energy, with a telltale hard *g.* I had recently learned the Japanese word for certain gates, *torii,* and I asked him if it could possibly have some link to the German *Tor.* It does not, he said—*torii* means abode of birds, and in both Europe and Japan the concept is too old; it predates the cultural and linguistic transmission lines across Eurasia erected by ships and now, of course, maintained by airplanes.

There's something about *gate* that lifts a place-name from the surrounding map, that embeds a name into the grain of a place, into the long miles and years of geography and history that surround it. Walking the cities of the world, or resting in a café, reading about a place overflown the night before, a pilot comes across many stately gate names, whether derived from the man-made or natural entranceways to cities: the implausibly ancient-looking gate of Fort Canning, in Singapore; the Golden Gate of Istanbul; or, indeed, the Golden Gate of San Francisco or the Lions Gate of Vancouver and the bridges named for them. I like the new bridge over Tokyo Bay near Haneda Airport almost as much as its name, the Tokyo Gate Bridge. In the mountains of Turkey stand the Syrian Gates, a high pass I've occasionally flown near, through which, a colleague once told me, Alexander the Great marched.

Travelers may find gates named not for where they stand, but for where they lead, to be particularly beguiling. The Brandenburg Gate in Berlin led to the city of Brandenburg an der Havel, from which the state of Brandenburg took its name. Fittingly, this name, which foreigners may most associate with the gate in the heart of Berlin, was recently bestowed on the German capital's new airport. On my first morning ever in Delhi, I saw the India Gate on a city map and jumped on the subway to go and visit it; then, once in the train, I nearly changed course when I saw signs for the Kashmere Gate. Sometimes the gate names spread to the surrounding neighborhood and endure long after the original gate has gone, such as with the Toranomon, or Tiger's Gate, in Tokyo, once the south gate of Edo Castle. London place-names hardly get better than Bishopsgate and Moorgate, which now survive not as structures but as the sonorous legacy of the Roman-era London Wall.

When I regularly flew to Paris, now and again I thought of gates, and airports, as I rode into the city through the Porte—the Gate—de la Chapelle. I like to imagine that this long heritage of gates gives a flicker of grandeur to the word as it is used today—merely numbered or lettered—in airports. Gates are exactly what these are—transitional spaces between the airplanes and the airport, that can be locked and opened. What we walk through to enter the modern city. America's once-great railway stations, the airports of a previous era, drew lofty comparisons to the gates of medieval towns. I don't speak Norwegian, which may be one reason that the word for customs—*toll*, which I read after flights to Oslo's beautiful and modern airport—so often evoked the sense of a formal passage through a city wall.

I have never flown the inaugural flight on a new route, but the traditional welcome given to an airplane on such a flight is a gate made by the airport fire brigade, who launch arches of white water for the airplane to taxi under, a ceremony whose components—water, an arch—feel archetypal, certainly much older and simpler than the fire engines that produce it might suggest. As for when we disembark from an airliner and wander into a terminal, it's perhaps too much to think of a portcullis rising or flags snapping in the wind under the gaze of elaborately dressed guards. Yet medieval travelers, too, would surely have been bleary-eyed and hungry. We forget that some of our greatest and most modern airports, those enormous glass-and-steel structures designed by the leading architects of our time, will one day look old-fashioned. We can't know now what nostalgia they will conjure, what romance for a former age of journeys and cities—our age of journeys and cities—they will someday embody.

Before we land, pilots must pass through gates in the sky. In physics at school I learned about the potential energy a bowling ball has at rest, high on a shelf, say, and the kinetic energy it has racing across a floor. An airplane at cruising altitude has plenty of both—height above the ground, speed through the air. Yet by the time it has parked at a gate thirty minutes later, effectively it has neither: it is not moving forward, and it has no stored-up height.

Pilots must reduce a plane's altitude in reference to other aircraft, obstructions on the ground, the published approaches to runways, and the instructions of air-traffic controllers. The plane's speed must comply with the requests of controllers, who are tasked with making the most effective use of their airspace and runways, and with more general speed limits; just as automotive speed limits drop when you approach a built-up area or town, in much of the world there are speed limits that apply to all aircraft below a certain altitude. Most importantly of all, a plane's speed must be neither too fast nor too slow at touchdown. The wings of a modern airliner are so efficient that the more typical problem is too much speed, not too little. The process of giving or trading away all this height and speed is aptly known as *energy management;* it is one of the more challenging tasks for the pilots of an airliner.

Sometimes I hear it said of a type of airplane that it is *slippery* or that it *can go down or slow down;* the implication is that it can only do one of these well at a time. This is a compliment, meaning that the airplane and its wings are well designed; it is also a warning that it can be harder to manage such a jet's energy.

In order to ensure that the energy is correct at touchdown, we back these requirements up into the sky, not in the information-age sense of making a copy of them, but in the physical sense of moving them earlier in time and place. At these fixed points in the approach we articulate—out loud, to each other—whether we are too high or too low, too fast or too slow. These points in the sky are known as gates: places we can pass through only under certain conditions. We may distinguish between *soft gates*—suggestions on the day that take account of the weather, our weight, the winds—and *hard gates* through which we must not pass unless the aircraft's energy, among other factors, is appropriate for the remaining distance to the runway.

Planes slow down before landing for the simple reason that a faster plane will use more runway length to stop, and runways are not infinitely long. But at a certain point a plane cannot fly any slower with wings sculpted for high and fast flight. Then we must spread the aircraft's wings, using *flaps* and *slats,* panels that extend and lower from the back and front of the wings—a capability that is also used at takeoff, though typically to a lesser extent than at landing. Expanded wings are bigger, more curved. They're much less efficient but allow the airplane to fly more slowly, an inefficiency that makes sense when it is time to take off or to land on a runway of limited length.

Wings that are not expanded are *clean;* the process of expanding them may be called *dirtying up.* A helpful controller will often tell pilots to fly their minimum clean speed; they mean for us to slow down, but there is no need to be inefficient yet. The 747's wings have seven configurations—one clean and six dirty. During the approach the expansion takes place in stages. Each stage lowers both the maximum and minimum speeds of the aircraft, and so as each stage completes we can slow down and initiate the next. The fourth of the dirty configurations is typically used for takeoff; the fifth- or sixth-dirtiest are used for landing.

The sight of our growing wings, often accompanied by a sensation of slowing that symbolizes all we must undo for our return, is one of the pleasures of flight that's largely reserved for passengers, who not only have the time to muse on the mechanical, carefully staged undoings of our height and speed, but also have a better view of the growing wing itself. It's worth asking, the next time you fly, for a window seat at the trailing edge of the wing, or just behind it. The pleasure of this view before landing could hardly be simpler: here are our wings, spreading for the act of return.

Such expanded wings are also one of the marvels of watching a plane land from the ground. If an airplane passes right over you before landing—perhaps in a traffic jam near an airport, or if you are inclined to have a picnic at a place where airplane lovers congregate for just this experience—the spread wing, the easily apparent nuts and bolts of the air-arms of our species, may be the most breathtaking thing about the moment; aside, of course, from the sight of something the size of a 747 in the air at all. Its huge curved flaps extended into the wind, its engines pressing against this new and self-created drag, it looks like the arriving bird it is—legs reaching forward, wings wide, poised for the moment to come.

*

I still like to look up at airplanes from the ground and to have a window seat when I fly as a passenger. Such moments remain a separate realm of experience, almost entirely distinct from the work in the cockpit. I'm generally most moved by flight not when I land an airplane myself but an hour after I've done so, when perhaps I'm on a freeway leaving the airport in Los Angeles and I see another airplane like mine only a few hundred feet over the ten streaming lanes of low traffic, the flash of the sunlit wings strobing over the cars. The forty-year-old me watches the landing airplane with a certain technical and aesthetic interest; the child in me can't believe that I've so recently been one of the two or three people who guided an airplane in its last thunderous moments over the upturned gazes of the five-year-olds of the world.

It's now about an hour before landing, six or seven hours before I'll walk through that temple gate in central Tokyo. We are plotting our passage through the gates of Tokyo's skies; we are planning how this airplane, which has climbed ever higher since we left the surface of the world near London, will descend and slow and return.

I make a few notes for my announcement to the passengers. I mouth through the words in Japanese before I speak. Near the end of *Citizen Kane*, Susan asks what time it is in New York. Eleven thirty, Kane tells her. Long-haul pilots may smile at her response: Night? Only Greenwich Mean Time is displayed in the cockpit; so as I always do, I check with a colleague to confirm my calculation of the local time at our destination. The changing language of place is reflected in the form such questions take in the cockpit. What is the time "here"? I ask, rather than "there," though we may still be 400 miles, a day's drive, even a time zone away from our destination. Here and there, in any case, are about to meet.

We tend to think of journeys, even air journeys, as lateral or curved endeavors, that we have moved over or around across the earth. But in the cockpit arrival has a much more vertical sensibility. At cruising altitude nearly all the world's complicated weather is below us. During descent we enter the weather of our destination not only from the side but from above. We return to *terrain,* the generic term for the earth's surface in the same way; we descend into the realm of mountains that may be beside us rather than below.

After we have *briefed the arrival*—the weather; our target speeds and altitudes at the various soft and hard gates, and our actions if we fail to meet these targets; the runway; our expected taxi route after landing, which is one of the more complicated parts of many arrivals—there are typically a few quiet minutes before we are given our first descent clearance.

When this clearance comes, we dial it into the autopilot, and at the appointed moment the engines roll to idle and the nose begins to drop. "Here we go," says the captain. You might think that this phrase would come at the start of a journey, when we push back from the terminal, and that certainly such momentum-conjuring invocations might come at the moment we begin the takeoff roll. But I hear myself say: "Here we go," most often near the end of a flight, at what's called the *top of descent,* the point we leave the high cruise. Here we go—down into the space that is different because it is lower, down to our destination that stands beneath it all.

Because runways rarely align with the direction of a journey, an airplane makes many of its largest and most dramatic turns right after takeoff, when it is taking up its route, and not long before landing, when it leaves its route to align with the runway. Narita Airport is not far from the coast, to the northeast of Tokyo itself.

On this bright morning the wind is from the north, so we make a clockwise series of turns around Narita, bringing us far to the south, over the water. This route takes us nearly over the airport, and below I see the exact spot where our airplane will reach the earth in fifteen minutes' time; we are pleased to see that our gate is unoccupied. Such direct overflights of the runway we are due to land on are a reminder of the extremity of our speed and altitude, of our still-mighty energy. There is no pulling over on the side of the road to let someone out; the only way to get to the place directly below us is to fly away from it at several hundred miles an hour.

In the last few miles before landing, airliners may follow a radio beam projected outward and upward from the runway. Pilots—or their autopilots—want to lock onto this beam, to follow it to the runway. Airliners generally approach the beam from a side angle, and because this angle, the wind and the speed of the airplane can all vary, the aircraft's final turn to capture the beam can be quite gentle. This morning, though, the wind is blowing us across the beam, so the autopilot reacts with a much sharper turn, to avoid the aircraft being blown through the beam and out the other side. Such a turn, the last major turn before landing, is worth watching for from the window seat. Its apparent certainty and vigor, as the airplane and pilots seize onto the course that leads to the runway and the flight's end, is as good an image as any of both aerial freedom and of the moment it is undone.

Once I flew from Moscow to London late in the evening. Our flight had been delayed by heavy snowfall in Moscow, and when we arrived in the skies over London we were the sole remaining airplane bound for Heathrow that night. It was a flawlessly clear night over the city. Though we were still outside the

M25, the ring road around London, we could already see the airport at the far side of that circle, across the city it encloses. We soon reported "visual with the runway," though we were still more than 25 miles away. On such clear and quiet nights there's no need to follow the beam through cloud or rain, no need for the complicated regime of speed controls that are typically applied to separate us from other inbound airliners at such normally busy airports.

"Very well," said the controller, when we reported we were visual. The instruction he was about to issue was a rare reminder to him and us of the smaller airports that both controllers and pilots train at. "You are cleared visual approach; *free speed;* all turns toward the airfield." It was nearly midnight when we sailed over the lights of the capital in the direction of the light carpet of the runway. And just this once I came to one of the world's busiest airports with a sense of freedom more familiar to pilots of a former age, returning from a late sortie to land on a grass strip, edged by a string of lanterns through the newly fallen darkness.

Most people I take into the flight simulator are charmed more by the experience of landing than of takeoff. Although at takeoff the runway dominates the windows, the destination is the vast sky above. Our eyes are drawn upward; they follow our intention as we move, in many senses, from the specific to the general. At landing this is reversed. The whole airplane, every mile of the flight, has been aiming at this country, this city and airport, but above all at this runway, a few miles northwest of Japan's Pacific coast.

The technologies that bring us to this point, to this sight of a city, still amaze me. We see the point from so far away. We see it, in essence, from the other side of the world, through fog and cloud and the skies of many countries; we see it not through the intervening rock but from far around the curve we will fly, from another day. Whenever I read that an example of tool use has been discovered in the natural world, some flicker of technology among another species, it suggests a continuum: from a sea otter pounding smartly with its rock to the airplanes that are guided across the world by the full light of our creations and in the last moments by our own eyes.

Some people who dislike flying specify that what concerns them is the feeling of not being in control. Another reason, I suspect, is that they cannot see their direction of travel. If it's not normal for humans to move so fast, it's even less natural to see only sideways while doing so. Even on a train the windows are big enough to show more of what lies ahead. Today, on a typical single-deck airliner, the cockpit occupies the position where the sides of the fuselage curve around to the nose, so it is only the pilots who can see forward. But on the double-decker 747 the cockpit is upstairs, so the two passenger seats nearest the front of the lower deck do in fact offer a partial forward view. Their two occupants will be able to see something of Japan straight ahead this morning, to watch our return to the land not only below the plane but also in front of it, to arrive here as simply as we do.

On most approaches we do not need to see the runway until the "DECIDE" call, in the last fifteen seconds or so of a flight, but usually we see it long before this, when the aircraft turns onto its final course or breaks from cloud. From far away a runway appears like a punctuation mark, a bracket tilting away along the ground.

At first it looks so small, marked off as precisely and preciously from its surroundings as a painting on the wall of a museum seen from far across a room.

When I'm first able to pick out the runway from the surrounding world, I may announce: "I've got it." Occasionally I hear colleagues announce: "Land ahoy," even if we have not been over the sea at all during the flight. But this is the most apt expression; from above, the edges of the runway mark off the only useful land in all the world. A few months before this flight to Tokyo I had landed in Vancouver, in an unexpected snow squall. For much of the approach there was no horizon to be seen, only the pattern of approach and runway lights hanging in the haze and tilting gradually toward us, as if we were sailing to the floating runway of a city in the clouds.

Many airports, like Narita, have multiple runways, and then the whole complex, lifting toward us, looks like a city itself, which the largest airports practically are. Approaching an airport with parallel runways, all sensibly aligned to the wind, is like approaching a city on an interstate that is getting busier and busier and suddenly noticing a barrier with, just beyond it, another set of lanes going in the same direction. Often passengers will see other aircraft paralleling their own plane's path, while on the wide roads below vast streams of lives and vehicles are heading toward a tower of clustering skyscrapers or to the same airport, all of us about to enter a city.

I now have a clear view of our assigned runway at Narita ahead. I remember, for an instant, the summer I came to Japan in school or the times I later flew here on business trips. I wonder who is among the passengers today, what songs they are listening to as they gaze out. "I am visual," I say. I disconnect the autopilot and silence the whooping siren that warns me I've done so. We lower the landing gear just before we cross the coastline. We complete the extension of the flaps and read the landing checklist. The air is bumpier now, yet another physical sensation that, like the spreading wing and the changing tune of the engines, marches hand in hand with the growing view of return.

Jonathan Livingston Seagull found that when he flew low over the water he could fly "longer, with less effort." Many pilots, whether or not that book inspired their choice of profession, will recognize just what he meant. When an airplane is in the final stages of the approach, a certain amount of power from the engines, paired with a certain angle of the nose, guides it down to the runway. But these settings must be slightly changed toward the end of the flight. The wing starts to produce more lift when it is near the ground, even if nothing else has changed. On the 747, I feel what is described as a *float* through the controls, a sudden resistance by the airplane to descend as willingly as before.

As the plane approaches the ground, the air beneath it can no longer move out of its way in time. So the air begins to act like a pillow. The proximity of the ground also prevents the vortices that spin off the wingtips from forming properly, which further enhances the efficiency of the wing. When an aircraft experiences this we say it is entering its own *ground effect*. The next time you see a 747 sailing over a park or a highway only moments before landing, at about the elevation of a twenty-story building, consider that this is the height at which the great restless jet begins to settle itself on the air you are breathing beneath it, a parting gift of antigravity from the sky or a welcome from the earth the airplane is coming home to.

At takeoff, particularly on a heavy airliner, there is an exquis-

itely balanced moment of hesitance when, at the stage called
rotation, the plane's nose is first raised. This sense of aerial equivo-
cation is not entirely imaginary, not entirely a consequence of any
lingering, atavistic disbelief in the possibility of ever taking to the
sky. At rotation the nose lifts, which means the tail must fall, and
passengers in the rear of the plane may correctly feel that they
have lowered in the sky at the moment they expected the opposite
to happen. They go down before they go up. Then at liftoff, which
comes shortly after rotation, the aircraft's weight at last leaves the
wheels and rests fully into the upturned arc of the more sharply
bending wings, which again may give the briefest sensation of settling back.

These effects, which give way to soaring almost before we can
recognize them, add up to flight's most liminal moment, as if the
articles of faith, or the numbers behind the physics of the enter-
prise, must be quickly incanted or calculated again each time we
take flight.

Takeoff's brief pause, that hanging in the in-between, finds its
twin at the end of each flight. In the last few hundred vertical feet
of our journey from London, as I start to feel the ground effect,
I lower the nose slightly, remove a touch of power. The captain
calls out the new thrust I've intuitively set so that I need not look
at the gauges. At about 30 feet above Japan I pull the nose up and
begin to close the thrust levers. I feel again that moment of poise:
the sense that continued flight is as likely as anything else, that a
question has been asked but not answered. Then the hard-won lift
runs like water from the wings, and we land.

Often I fly over a place that is tied to my own life in some way.
Sometimes when I fly to Boston I don't stay at the official crew

hotel; instead I visit friends north of the city. The next evening the climbing airplane passes right over their town. If I see the river near my friends' home, I will think about the table they laid for me and the grateful pilot who came to their place and felt no sort of lag, until it was time to fly away.

Sometimes I fly over a place I have known in another context or time, and the sight gives a new life to the memory, a depth I might not find even if I traveled there again. When I was a child my family spent a few summers on Lake Winnipesaukee, in New Hampshire, in a cabin where even in July the mornings were cold enough to require a fire. Occasionally I see the corner we knew of that lake from the sky. To come across it again now that I am three or four times the age I was when I last swam there, in seasons other than the summers I knew it in—to see it frozen and snow-covered, or lined with autumn colors, looking from high up as if the turning trees were mere red lichen around a pool lying in the indent of a rock in a forest—has been a happy experience. In summer, when I see the boats on the lake below, their wakes like the trails of comets on the sky-blue water, and think about the young families in them, it's not quite that I feel that I am looking back in time. But certainly from above, and from so many years later, the lake takes on a wholeness that is indistinguishable from my memory of it.

Often I hear colleagues say, in a jet over Britain, that they are within a few miles of their own house—or right over it. They say this without looking out the window; sometimes they say it when we are in cloud. They know the bearings and the beacons, the miles to home.

On flights from London to Mexico City I now and then pass over the part of the world that I know best, western Massachusetts, where I grew up. We always spent all the holidays with a group of three other families who were friends of my parents; they are like aunts and uncles, and their children are like my cousins, all the more so now that my parents are gone. Western Massachusetts from above looks like the place they came to and the place I came from. There is not much to distinguish it from the surrounding forested lands. I'm comforted, somehow, that most everyone else onboard would see only trees.

Even the mountain here, Mount Greylock, at only 3,491 feet is hard to pick out, though it is the tallest in Massachusetts. The war memorial on top of it, a tall stone tower beneath which we had many picnics when I was a child, is the best clue. When I find the mountain I think of Herman Melville, looking up at it from his desk in Pittsfield, between thoughts of less landlocked places. Often I overfly western Massachusetts not long after crossing the ocean from Europe. If there is a solid deck of fresh snow or cloud over the land, I remember that it was winter's obliteration of this countryside that gave Melville "a sort of sea-feeling here in the country," that he would look from his house on the land as from "a port-hole of a ship in the Atlantic" and wonder, as the winter wind howled around, if there was "too much sail on the house."

Before I became a pilot, if you had asked me to talk about a city that I had visited, I might have thought first of its architecture, its food, or a memorable event from my first visit there. Now I tend to think first of its geographic situation: what it looks like from above and far away; whether it is on the edge of mountains or the sea or a desert; what ideas of land give way, like distance itself, to the fact of Vancouver or Milan. These are places that feel different to me even when I walk through them, because I know what it looks like to arrive in them from the sky. This is one of the satisfactions of my job that surprised me: not flight itself but this almost anachronistically literal awareness of how cities rest on the physical world.

There is another category of city, though, for which the aerial, geographic sense of a place does not augment other impressions of it, because I have no other impressions. Doha, Athens, Kiev, Ankara, Tripoli, Buenos Aires, Zagreb; I have landed in these cities and then flown away, without ever leaving the airport. Sometimes I have not even left my seat.

In this category of cities it's Moscow that I've flown to most often. I could tell you how unusually round Moscow looks, the metropolitan phenotype that is the privilege of cities born in flat and landlocked places. I might mention Moscow's multiple, concentric ring roads—one of which roughly corresponds to the city's medieval boundaries and gates—that glow in the pitch-black winter nights like the rings of an electric cooktop. When I flew the Airbus and went often to Moscow we were not permitted to fly over the city center, nor were we usually permitted to fly around it in a counterclockwise direction, and so we would fly nearly three-quarters of a circle around the city, as if it were an aerial traffic circle. On such arrivals it felt that we were orbiting, caught in the gravity of the city, and that the aircraft's wide, long turns echoed the purpose and the shape of the ring roads below.

I could tell you more than I ever expected to know about Moscow's weather, and something about the Muscovites I met who worked at the airport or whom I met on my flights. From above at night, I have seen the whole of the city more clearly than many people who live there ever will, set on the land like some great fired wheel turning on the snow, encircled by the dark forests, under the navigation lights of the airplanes banking around it.

Yet in nearly every other sense I am a stranger to Moscow, and perhaps the worst kind, who may decide he knows something of a place from only a series of brief exposures to the most abstracted and antiseptic of views. What lie within the ring roads, for me, are lights, not individuals. Whatever I might imagine of the lives in the city comes from television and novels and history books.

Perhaps this is only an extreme version of how we experience any place, even one where we get out and walk around, even the one we live in. We will never know more than an absurdly small portion of any city or landscape. But still, when I'm asked if I have been to Moscow, the question makes me a little uncomfortable. No matter how many people I've taken home to the city, no matter how many times I have followed its transformation from a distant glow to a circular galaxy of light and at last into the physical sensation of touchdown, I feel that no is the only possible answer to the question of whether I have been there.

The skies of Alaska are relatively busy. There are many airplanes and aviation is important, for good reason—with its residents living in a few concentrated areas and many small settlements separated by vast distances, towering mountains, inhospitable terrain, and water in its least convenient forms, Alaska is a microcosm of the jet-age planet. John McPhee, in *Coming into the Country*, describes how Alaskans, if asked whether they've been to a place that they've seen from the air but where they haven't stood upon the ground, may reply with a qualified sort of yes; they may say that they've "flown it."

The question of what it means to fly a place arises not only for cities but for whole lands. I have long been fascinated by Arabia— by its appearance on maps and globes or in old tales encountered in childhood, by the name I saw so long ago, etched on the side of a plane moving slowly over the ice-ridged taxiways at Kennedy Airport. When I fly now over Arabia and imagine and say its names to myself—Jeddah, Medina, Mecca, Dhahran, and Riyadh—and then see something of Saudi Arabia's present day, its solar panels and crop circles, the cold, sprawling glitter of desert cities in the furnace of summer nights, the coasts and highways shining up like the most perfect map of the country in my mind, I feel that I can say I know something of the place.

But airplanes, as I first began to realize when I marveled at that Saudi jet, connect our ideas of place as much as the actual places. From above, it is hard to imagine learning something about Arabia that could not be instantly incorporated into my sense of it from the sky, that could not fit into an aerial viewfinder whose breadth is both its marvel and its weakness. My skyborne sense of a place like Arabia could hold almost anything I half remember hearing about it in childhood or anything I might learn about it elsewhere; and so, I fear, maybe it holds nothing at all.

Sadly, perhaps, this feeling did not change as much as I thought it might, when I went to Riyadh for the first time not long ago. I slept for most of my short visit and made my way out of the hotel only on two brief excursions. When you hear a song you know well on the radio, at low volume in a place crowded with other noises, you can just about follow it; but the unfamiliar song that comes next is something you cannot begin to get traction or purchase on—you only hear the occasional deepest beats. It's almost just noise. That's a feeling I associate with my briefest stays in places I don't otherwise know well, such as Riyadh; that even having spent a night in the city I have still only flown the place.

After many years of flying over Greenland—flying the place that I perhaps most love to fly—I was given a book by Gretel

Ehrlich about it. She narrates the story of Ikuo Oshima, who moved from Japan to Greenland many years ago and adopted a traditional hunting lifestyle in Siorapaluk, one of the northernmost settlements of the world. From his new home, Oshima sees satellites pass overhead—satellites, he is told, that can read car license plates. He wonders about one of them, looking now down at Tokyo, then, hardly later at all, down at him in Greenland, "standing at the ice edge dressed in polar bear pants and holding a harpoon." He wonders how the satellite feels. "Maybe confused and broken." More than once I've flown over Japan and Greenland in the same week. The satellite, I suspect, has place lag.

At the end of *Stuart Little*, Stuart comes to a fork in the road. Unsure of where to go, he stops, then meets a telephone repairman working nearby. The repairman advises Stuart to head north and tells him of some places—orchards, lakes, fields bordered by "crooked fences broken by years of standing still"—that Stuart may find in his adventure to come. "They are a long way from here," warns the man. "And a person who is looking for something doesn't travel very fast."

When I spin a globe my inclination is to stop it at Mongolia. Have I been to Mongolia? I would say that I have flown it, and very fast. The border of Mongolia comes not long after the Siberian city of Novosibirsk. Then I may see the perfect blue computer-screen circle that stands for the airport named after Genghis Khan. I might have dedicated my whole career to a place such as Mongolia, the name that caught my ear or eye as a child. I might have put my whole life into some subspecialty of its history or geology or linguistics, lived there even. But I long first of all for the name, then the place it attaches to, a place that is pleasing to me to imagine. From above, perhaps, I see this imagined place as much as I do the real land below.

Sometimes when I see the first peaks of Mongolia I think of Stuart, stopping to consider a stranger's advice, then looking over "the great land that stretched before him" and driving off into the morning. The reality of the place and the morning ahead is indisputable. There is no doubt, this is Mongolia that rolls into view as ordinarily as the day. Again and again I have seen the sun rise on the place itself, on peaks dusted with snow in all seasons, and seen the light descend into the shadowed glow of tawny valleys, where it falls across the occasional surprise of a road. Then, whatever the truth of the place, whatever I've gained or lost, the whole of it turns away under the line of the wing. The great eye of the world blinks, and now we are somewhere else.

Whenever I am not sure where I am in anything but the most literal, the most what-is-the-name-of-this-city sense, whenever I have to stop and think what continent I was on a few days ago, I try to remember that flying has deepened my love of home.

I don't mean by this that I appreciate more than I otherwise might the specific advantages of the places where I grew up or have lived. I do appreciate some things about home more but others less. Seeing so much of the world, a pilot can too easily wonder why every city can't have the train stations of Beijing, the outdoor swimming pools of Helsinki, the cycle paths of Amsterdam, the friendly taxi drivers of Vancouver, or the devotion to public greenery of Singapore.

Rather, the deepening gratitude I feel for home relates to it as the place that, wherever I am flying, I know I will return to and be still. Home is where my sense of place can find itself most easily, where my rooted half, the one that refuses to travel at anything faster than a brisk walk, knows to wait, when he wakes up and realizes his peripatetic companion has taken the passport from the desk drawer and gone away again.

To come home from a trip to a high place and a far city, from hours over the tundra or distant oceans, is a sudden and joyful deceleration. I feel this almost physically. As the airplane slows on the runway, both the actual speed and the place streaking, the self-blurring, begin to end. And once home it is the simplicity of the ordinary things, rather than the shock of difference, that is heightened by the scale of the journey. All those miles, all those hours over ice or sand or water, to return to a snack taken from a cupboard, to a photograph on a shelf, to the closet quietly closed with the suitcase at last at rest inside. I may go out to eat with friends, comforted that they do not know what other seasons and countries I have seen since we last ate together. I like that they rarely ask where I have been, as if I have not been away at all, as if that other self had stayed at home or walked across town to be with them.

Sometimes, if I have no errands or plans, I will not leave home for a day or more after a long trip. There's a new pleasure to the confinement of the rooms and in the weight and simplicity of the small tasks and details, a wonder they reacquire in direct proportion to the time and miles I have traveled, a quality they accrue only because I left. If a hint of place lag is experienced even on returning home, perhaps such days of self-mooring are a way to minimize further confusion, of not changing once again the place—not even to a coffee shop down the street—where I find myself.

If I go out at all, it might be to a park, where the sight of actual soil after so many earth-miles is so striking, and my slowness above it feels like a kind of miracle and another unexpected gift of the airplane's speed—as on that flight in a small plane over England years ago, when the fighter jets passed us so quickly that our newly evident slowness suddenly became a gift. This, too, is place lag, but when it's induced by a return home it is not an uncomfortable feeling. Until I fully inhabit my place and time again, it's as if I can still see what my life looks like from a certain distance or altitude—an appreciation made clearer, perhaps, by listening to a song I happened to have heard while I was away. This is how the closely spaced atoms of home come to fill the horizon when I return, as simply as the whole world came to fill the cockpit windows when I left.

It hardly matters where we go, only that we've been away. An old friend of my parents is from Wisconsin but has lived for a long time in my native part of New England. I tell her that when I come back to New England, the hills look right, that no matter where I am, they are what I am always expecting to see standing beyond the end of every road or rising from the far shore of every lake. She laughs and says that the longer she is away, the more right Wisconsin looks when she returns to it. She flies from her hilly current home to her flat first one. She watches the land change beneath her. She lands, leaves the airport, and drives out to the farm she grew up on. Something opens, she says.

Airline crews come to know a life of motion, of transiting the physical miles between our memories or ideas of places. But this is only a more extreme version of what every air passenger experiences when they make a new journey or retrace one they—or their parents or grandparents—made long ago. Such travels are emblematic of our age: of globalization, urbanization, immigration. In such a world, time becomes indistinguishable from geography, not only in the original sense of the slow migrations of continents or of the arcs of time zones that we have drawn on the turning planet, but in the motions of our own lives and families. When that friend of my parents returns to Wisconsin, it's unnecessary, she tells me, to even try to tease out the years she's been away from the miles of her journey. She flies back in time.

At its best, this experience is a wonder. But the airplane gives airline crews, at least, so many places that the effect is all but permanently disconcerting. On Friday I fly over Iran, closing on the Turkish border near the salt lake of Urmia. As if in mimicry of the sky above it, the lake's deep-blue center fades out to tan edges. On Monday I am over Utah and I blink twice at the Great Salt Lake below. The feeling you may have on the first cold day of winter, when you put on a heavy coat you have not worn in eight months and your fingers encounter a restaurant receipt in the pocket, from a meal you barely remember—there is a geographic, planetary equivalent of such a temporal or seasonal disjointedness, and I experience it almost constantly. My wallet is a library of subway cards. I find coins from Kuwait in my pocket and I cannot begin to remember when I was last there; I take out a pair of shorts from the closet at home, shake out the sand in the pockets, but I cannot name the beach. Often I cannot name the sea.

A few times it has happened that I've passed by the sign for a restaurant near where I grew up in Massachusetts, named for Bombay, the old name of Mumbai, and then a day or two later have flown to India, to the city itself. Later, walking the rooms of the house where Gandhi stayed, or passing the city's famous outdoor laundry, or staring out from an auto rickshaw at the tectonics of a Mumbai traffic jam, I have thought of that restaurant in the snow not two days earlier, as bewildering as a barely remembered dream. On the rare occasions when I have not enjoyed my job, it is because I've felt that I am not in this place or that place; I've felt I am nowhere at all.

One December, however, the Bombays, the twinned Mumbais, ran in reverse. I started my day early, in the city in India, flying to London as a pilot and then on to Boston as a passenger. Then I drove west, toward the town where the families with whom I had always spent the Christmas holidays since I was born were already gathering.

Not long before I passed the Indian restaurant, as the winter night gathered over the Berkshires, the year's first snow began to fall. The world reassembled as the place that means the most to me, as the small place that home can only be. When I saw the sign for the restaurant the thought of the various Bombays, the scatterings from the airplane's prism, was not troubling. It was something to hold and marvel at, and then set down.

Every landing is a return from the possibility of all places to the certainty and perhaps the love of one. Many years ago I flew to Toronto as a passenger. It was an overnight flight, and the sun rose before we started our descent. It was summer, the whole window of morning was filled with blue and green, and I was listening to my music as Canada rose to meet its day. When the airplane was on its final approach, a quick shadow, moving in the distance, caught my eye. It tracked effortlessly over forests and ponds and along the lanes of the highways. Eventually I understood it was the shadow of the airplane I was in: an eclipse of the earth, made in the image of the airliner returning me to it.

I tried to imagine where the early sun must be, far above and on the other side of the craft, to throw this shadow onto the earth. I stared, rewound the song in my headphones, turned up the volume, and grabbed my camera. I had never seen this before and I thought it might soon disappear. But the shadow stayed roughly in the same place in my window. It grew in size as the world hastened under it. I realized that the airplane and its growing shadow were approaching each other. After so many parted miles they would meet again at the moment of touchdown.

Only a few times since that morning have I seen the sun and the journey's end aligned so well that the shadow of the jet appears and trembles on the land, as if in anticipation; as if the sound of the engines or the growing form in the sky has helped the shadow to remember what first cast it. Here is land, both noun and verb. Here we are, coming home.

The shadow keeps perfect pace with the widening wings. It crosses the earth as simply as the plane, as simply as our eye, or as if it was a kind of light, the mark made on the planet by our own falling gaze.

I look out. Each time I see this growing shadow it makes me smile, and I can almost believe it's the first time I've been in a plane. There it is now, I think, there it is, as I turn up my music again. Through the gap between the seat and the wall ahead of me I realize that another passenger has seen it, too. She looks back at me, points to the outside. I nod and smile. We both lean forward and turn, pressing against our seat belts, to watch.

Acknowledgments

The world of aviation is as wide and varied as the planet. In the course of researching and writing I have become only more aware that many pilots—particularly those of smaller planes, in smaller airlines, based at smaller airports, or outside Britain—will have experiences that may be very different from those typical to my corner of it, a limited realm that I have tried to describe as accurately as I can.

I would first of all like to thank all of my colleagues on the flight deck and in the cabin—and in turn, our many colleagues on the ground, without whom no flight would ever push back and upon whom our safety depends in a thousand unheralded ways—for their enthusiasm and professionalism, and for all they have taught me about airplanes and the world. I have no reason to doubt what several retired pilots have reported to me—that it's not only the flying, but those who share our love of it, that make this the best job in the world. I am also grateful for the cadet sponsorship program in which I enrolled in 2001. Such programs open the profession to those who could not otherwise afford it and are more necessary than ever.

I would also like to thank those who first taught me how to

fly—my instructors and the staff at what is now called CAE Oxford Aviation Academy. I'm very grateful to my colleagues on training course AP211—Jez, Bomber, Seb, Cat, Neil, DAVE!, Adrian, Adam, Kirsten, Chris, Balbir, Lindsay Boy, Lindsay Girl, Mo, Hailey, Carwyn, and James—for all their support ("You'll be fine, mate!") and friendship on the course, and since. Whenever I hear one of you on frequency, speaking to Gander Radio or Iran Air Defense or the Heathrow Director, I'm reminded how much I wish we could fly together more often. Thanks also to Simon Braithwaite for his company in Cape Town, and to Nigel Butterworth for inviting me to the cockpit on the way from Narita to Heathrow, way back when you could.

Several individuals read the entire manuscript and offered their thoughtful comments on it. I'm grateful to Mark R. Jones and Kirun Kapur (for their formative thoughts not only on the "Night" chapter, the first one I wrote, but on all the others since), Steven Hillion (who first read a complete early draft and guided many of my subsequent revisions), Desirae Scooler, Harriet Powney (whose attention to detail was a particular inspiration), Cole Stangler, Don MacGillis, Sebastien Stouffs, Douglas Wood, John Pettit, Ian Slight, Tony Cane, Mary Chamberlain, and Alex Fisher (whose knowledge of both the technology and history of aviation is like that of no one else I've ever met).

Various experts and colleagues offered their advice on particular chapters. I am grateful to Emma Bossom of the Royal Aeronautical Society, Richard Toomer and David Smith of the British Airline Pilots Association, Paul Tacon of the Honourable Company of Air Pilots, Marc Birtel and Shaniqua Manning Muhammad of Boeing, Mike Steer of CAE Oxford Aviation Academy, Peter Chapman-Andrews of the Royal Institute of Navigation, and Paul Danehy of the NASA Langley Research Center for putting me in touch with the following experts.

In writing the "Place" chapter, my colleague on the ground, Mark Blaxland-Kay, never tired of my questions about navigation and route planning. David Broughton, Charles Volk, Larry Vallot, Andrew Lovett, and Brian Thrussell cheerfully answered my questions about inertial navigation and magnetism. Nanda Geelvink, Brendan Kelly, Mireille Roman, and Robin Hickson offered welcome assistance on waypoint names and airspace structures.

The "Air" chapter benefited greatly from the patient technical assistance of Jennifer Inman, Matthew Inman, Andrew Lovett, Brian Thrussell, Stephen Francis (who first introduced me and my course mates to some of this material), Stuart Dawson, Eugene Morelli, and R. John Hansman. Dave Jesse and James H. Doty offered their assistance with the section on radio altimetry.

Douglas Segar gave me a great many invaluable comments and suggestions for the "Water" chapter. Jeff Kanipe, Stephen Schneider, and George Greenstein offered their kind assistance with the "Night" chapter. Their comments and evident enthusiasm for the night sky gave me a hint of my father's regret at not becoming an astronomer.

Helen Yanacopulos, Jamie Cash, Eleanor O'Keeffe, Ulrike Dadachanji, Mark Feuerstein, Martin Fendt, Terry Kraus, Amanda Palmer, Vinod Patel, Dick Hughes, Pamela Tvrdy-Cleary, Julia Sands, Karen Marais, Chris Goater, Haldane Dodd, Anthony Concil, Mitch Preston, Drew Tagliabue, Mark P. Jones, Hilda Woolf, Mei Shibata, John Edward Huth, Al Bridger, Kannan Jagannathan, and Wako Tawa offered their kind assistance with other sections of the book.

While I am grateful for the assistance of friends, colleagues, and experts, any remaining errors are, of course, my own.

In addition to the pilot colleagues and trainers I consulted and the manuals, training materials, and charts to which I am grateful I had access, I relied on several other written sources. *The Airplane in American Culture* (edited by Dominick A. Pisano) and Joseph Sutter's *747: Creating the World's First Jumbo Jet and Other Adventures from a Life in Aviation* were both fascinating and useful. Two textbooks of the American Meteorological Society, *Ocean Studies: Introduction to Oceanography* (3rd edition) and *Weather Studies: Introduction to Atmospheric Science* (5th edition), were helpful in writing the "Water" and "Air" chapters. The well-named *An Ocean of Air* by Gabrielle Walker was both captivating and instructive. *Introduction to Avionics Systems* (3rd edition) by R. P. G. Collinson was a resource for various sections. John Huth's *The Lost Art of Finding Our Way* was an excellent and helpful survey of the history and future of navigation.

It has been a great pleasure over the last eighteen months to walk into offices filled with books and those who love them. I am grateful to my agent, Caroline Michel, for reaching out to me and for all her subsequent support and encouragement, and that of her kind colleagues. My editors, Clara Farmer and Susannah Otter at Chatto & Windus, and Dan Frank and Betsy Sallee at Knopf, and their colleagues on both sides of the ocean, particularly Maggie Southard, Sara Eagle, Gabrielle Brooks, Lisa Gooding, and Vicki Watson offered clear-eyed, patient, and warm-hearted guidance and support throughout this project.

Finally, I'd like to express my thanks to my family and friends for their love and support throughout this project and for getting me away from it from time to time. Thanks to Kirun for understanding about music and the window seats of all sorts of vehicles;

to Nancy, for her love and support throughout the years and for always encouraging my interest in flying; and to Mark, for everything. Silas, Anjali (who reminded me about the one ocean), and Lola—I hope that someday you can visit the cockpit during a flight, as kids could in the old days.

Permissions Acknowledgments

Grateful acknowledgment is made to the following for permission to reprint previously published material:

Ann Rittenberg Literary Agency, Inc.: Excerpt from *The Artist's Voice: Talks with Seventeen Modern Artists* by Katharine Kuh, copyright © 1960, 1961, 1962 by Katharine Kuh, copyright renewed 1988, 1990 by Katharine Kuh. Reprinted by permission of the Ann Rittenberg Literary Agency, Inc., on behalf of Avis Berman, Literary Executor of the Estate of Katherine Kuh.

Curtis Brown, Ltd.: Excerpt from "Folklore of the Air" by William Faulkner (*The American Mercury,* November 1935). Reprinted by permission of Curtis Brown, Ltd.

Farrar, Straus and Giroux, LLC: Excerpt from *Gilead: A Novel* by Marilynne Robinson, copyright © 2004 by Marilynne Robinson; excerpt from "Part I" from "Midsummer" from *The Poetry of Derek Walcott 1948-2014* by Derek Walcott, selected by Glyn Maxwell, copyright © 2014 by Derek Walcott. Reprinted by permission of Farrar, Straus and Giroux, LLC.